How To Raise Poultry For Pleasure and Profit
The Breeding, Rearing and General Management of all Kinds of Poultry

by William M. Lewis

with an introduction by Jackson Chambers

This work contains material that was originally published in 1895.

This publication is within the Public Domain.

*This edition is reprinted for educational purposes
and in accordance with all applicable Federal Laws.*

Introduction Copyright 2018 by Jackson Chambers

The World's Largest Selection of Vintage Poultry Books

www.VintagePoultry.com

Self Reliance Books

Get more historic titles on animal and stock breeding, gardening and old fashioned skills by visiting us at:

http://selfreliancebooks.blogspot.com/

Introduction

I am pleased to present yet another title on Poultry.

The work is in the Public Domain and is re-printed here in accordance with Federal Laws.

As with all reprinted books of this age that are intended to perfectly reproduce the original edition, considerable pains and effort had to be undertaken to correct fading and sometimes outright damage to existing proofs of this title. At times, this task is quite monumental, requiring an almost total "rebuilding" of some pages from digital proofs of multiple copies. Despite this, imperfections still sometimes exist in the final proof and may detract from the visual appearance of the text.

I hope you enjoy reading this book as much as I enjoyed making it available to readers again.

Jackson Chambers

FOWLS—THEIR GENERAL MANAGEMENT.

There is a great diversity of opinion in regard to the management of fowls, the particular and desirable breeds for all purposes, &c. First of all, their

PROPER CARE AND KEEPING

is essential to success, for a person may have the best known breeds, and if they are not properly cared for they will, in nine cases out of ten, prove a failure. Therefore we wish to impress upon the mind of the breeder, in the outset, that this needs attention more than purity of breed or superiority of kind. As a friend of ours said, "there exists gross neglect of the poor birds generally." This neglect is not confined to persons who have no fancy for fine poultry, but extends even to many who have the reputation of being fowl fanciers. Still, as before stated, for poultry to be remunerative there *must* be good management. In

STARTING OUT IN THE BUSINESS,

plans should be well matured and digested before hand. A good, convenient poultry house should be properly constructed, sufficiently large to contain the number of birds one desires, warm and dry in the winter, well ventilated, and it should be kept scrupulously clean. The house should not be over-crowded, but just large enough. Nothing is made by over-crowding the hennery; on the contrary, it will prove detrimental. The fowls must be fed regularly and at stated periods. They must have plenty of pure water at hand at all times — this is of as much importance to the health of the brood as proper food. If possible, they should also be given, in addition, a plat of grass for a run. Place within the hennery a dust heap; this may consist of wood or coal ashes, sand, or dust from the streets. It should be kept under cover, so that it will not become drenched with rain or snow, and to it the fowls should have access at all times, to dust, and thereby rid themselves, in a great degree, of the numerous parasites which infest them. The habit of

GIVING TOO MUCH FOOD,

to poultry, in a short space of time, is a very bad one. If one notices their habits he will perceive that the process of picking up their food under ordinary, or what we may call the natural condition, is a very slow one. Grain by grain is the meal taken, and with the aggregate no small amount of sand, pebbles, and the like, all of which, passing into the crop, assist digestion greatly. But in the "hen-wife's" mode of feeding poultry, a great heap is thrown down, and the birds are allowed to "peg away" at such a rate that their crop is filled too rapidly, and the process of assimilation is slow, painful and incomplete. No wonder that so many cases of choked craw are met with under this treatment. Many other diseases which affect chickens might be prevented by breeders, were a little precaution taken in the simple matter of feeding.

TO PRODUCE EGGS.

More eggs can perhaps be obtained from hens by mixing breeds than by any other mode; and it is generally conceded that crossing also promotes the health of fowls far more than the vile practice, as some are pleased to term it, of in-and-in breeding. Little trouble need be apprehended from roup, gapes, cholera, and other diseases in poultry, if that care is observed in breeding and crossing that is so essential to all well regulated poultry yards.

POSITION OF THE HENNERY AND RUNWAYS.

As we said before, the hennery should be placed in a warm, dry location — (not in a damp, out-of-the-way place) — with runways ample to allow of plenty of exercise. Above all, care should be taken that vermin do not get a foothold in the hennery; for if they once make their appearance, it is difficult to exterminate them, and before the breeder is aware of it, his flock is over-run with them. Let the hennery be thoroughly cleansed with lime, (whitewash put on hot,) as often as once a month. If any of the fowls show symptoms of disease — which is frequently the case when in confinement — see that they are removed at once from the flock. Give good, wholesome food, with plenty of clean water; have the laying boxes cleaned and renewed frequently with straw, hay or shavings, and, with the help of the good housewife and children, there need be no fear of failure to profitably raise poultry. If one does not succeed in the first undertaking, he should not become disheartened, but persist in his endeavors to find out the cause of failure, and obviate it in the future.

MR. LELAND'S EXPERIENCE.

In this connection we give the reply of Mr. WARREN LELAND, Rye, N. Y., an experienced and extensive breeder of fowls, to inquiries from a gentleman who desired to engage in the poultry business in his old age. Mr. LELAND says:—"I have found that for every hundred fowls you must give up at least an acre. But rough land is as good as any. Hens naturally love the bush, and I lop young trees, but leave a shred by which they live a year or more. These form hiding places and retreats for them. In such places they prefer to lay. I have great success, and it depends on three or four rules, by observing which I believe a good living can be made by hens and turkeys. I give my fowls great range. Eighteen acres belong to them exclusively. Then the broods have the range of another big lot, and the turkeys go half a mile or more from the house. The eighteen acres of poultry-yard is rough land, of little use for tillage. It has a pond in it, and many rocks, and bushes, and weeds, and sandy places, and ash heaps, and lime, and bones, and grass, and a place which I plow up to give them worms.

"When a hen has set, I take her box, throw out the straw and earth, let it be out in the sun and rain a few days, and give it a good coat of whitewash on both sides. In winter, when it is very cold, I have an old stove in their house, and keep the warmth above freezing. There is also an open fire-place where I build a fire in cool, wet days. They dry themselves, and when the fire goes out there is a bed of ashes for them to wallow in. Summer and winter my hens have all the lime, ashes and sand they want. Another reason why I have such luck is because my poultry yards receive all the scraps from the Metropolitan Hotel. Egg making is no easy work, and hens will not do much of it without high feed. They need just what a man who works requires — wheat bread and meat. Even when wheat costs two dollars I believe in feeding it to hens. As to breeds, I prefer the Brahmas, light and dark. I change roosters every spring, and a man on the farm has no other duty than to take care of my poultry. I frequently turn off three thousand spring chickens in a single season."

BREEDING AND MATING.

Too many fanciers and farmers, otherwise earnest in their business, are very careless concerning their fowls. Interbreeding certainly degenerates — particularly when so promiscuously permitted in a flock of fowls as is common. There are the same good reasons for

MAKING CHOICE OF THE BEST BREEDS OF FOWLS

as for making the same choice in other stocks. For while a prime breed is as easily reared, fed and housed as a poorer one, there is a decided difference in the returns in favor of the former. If properly cared for, we do not hesitate to say that fowls of superior order do yield the farmer, even, the largest interest for the outlay he makes of any other stock he keeps.

In giving our own, and the opinions of others on the general principles of breeding and mating fowls, it will not be out of place to give here an illustration and description of

THE DIFFERENT POINTS OF A FOWL,

so that the reader may be able to name them, and judge therefrom, in his selection of stock for breeding purposes:—A, Neck-hackle; B, Saddle-hackle; C, Tail; D, Breast; E, Upper Wing Coverts; F, Lower Wing Coverts; G, Primary Quills; H, Thighs; I, Legs; K, Comb; L, Wattles; M, Ear Lobe.

SELECTION OF COCKS AND HENS FOR BREEDING PURPOSES.

A desirable thing in breeding is the selection of the cock. This, as all should understand, is a very important matter to be looked after; another is the proper proportion of hens to be given to the cock. To breed a good fowl of any kind requires thought, skill, observation and study. The cock in all cases should be of good size, perfectly healthy and vigorous; carry his head high, and have a quick, animated look, a strong and shrill voice; the bill thick and short, the comb of a fire red, bright color; a membraneous wattle of a large size, and in color resembling the comb. He should be broad-breasted, with strong wings; the plumage dark, the thighs muscular, and spry and trim on his legs; free in his motions; crow often, and scratch the earth with constancy in search of worms, not so much for himself as his mates; when he is brisk, spirited, ardent and clever in caressing them, quick in defending them, attentive in soliciting them to eat, in keeping them together in the day, and assembling them at night, he will prove as a general thing, just the bird to breed from. The good qualities of hens, whether intended for laying or breeding, are of no less importance than those of the cock. The hen is deservedly the acknowledged pattern of maternal love. When her passion of philoprogenitiveness is disappointed by the failure or separation of her own brood, she will either go on sitting, till her natural powers fail, or she will violently kidnap the young of another fowl, and insist upon adopting them. But all hens are not alike. They have their little whims and fancies, likes and dislikes, as capricious and unaccountable as those of other females. Some are gentle in their manners and disposition, others are sanguinary; some are lazy, others energetic almost to insanity. To succeed in the matter of the selection of hens for mating and breeding purposes requires care, study and a considerable degree of patience.

THE NUMBER OF HENS TO A COCK, ETC.

We have no hesitancy in recommending to breeders the following ratio of hens to a cock of the breed named:—Houdans, twenty hens to two cocks; Creve-Cœurs, eight hens to one cock; Buff Cochins, twenty-four hens to two cocks; Gray Dorkings, ten hens to one cock; White Leghorns, fourteen hens to one cock; Spanish, twelve hens to one cock; Brahmas, twelve hens to one cock; Hamburgs, fourteen hens to one cock; Polands, twelve hens to one cock; Game, ten hens to one cock. With this proportion of hens to a cock the vitality of the eggs will prove good, and at least eleven out of twelve eggs set will produce "chicks."

For breeding purposes, we inclose in a yard ten or fifteen hens of each variety we desire to propagate, and with them one cock; if we have two or more cocks whose qualities are equal, we think it preferable to change every two days, leaving only one cock with the hens at a time. Two weeks are necessary to procure full bloods, and we prefer the eggs the third rather than the second week.

We are told by a breeder of some considerable experience with fowls that to determine the exact proportion of cocks and hens to be allowed to run together for breeding purposes is not an easy problem. He says: "While with some varieties, as the Cochins, three or four, or even two, are ample, (though we have seen cocks of that variety that would serve ten or a dozen;) in others, twelve to fifteen are not too many. It is impossible to give any definite number for a rule. We have had pairs that did well, the eggs hatched well, and the hen did not suffer from the over-attention of the cock; and again, we have been obliged to put in one, two, three, four and even more additional hens of common stock, with a trio of pure-bred fowls, to keep the blooded hens from being injured. Especially is this the case with the Houdans and Creve-Cœurs; the cocks of both these breeds seem to be very vigorous, and require not less than four or six hens to run with them. The Dark Brahmas also need not less than four hens with the cock when he is young and vigorous. It was a favorite theory of ours, some years since, that poultry should be bred in pairs or trios. Because in the *wild* state, they ran in pairs, so also, should they do in the domesticated state. It is needless to say that our theory would not work when carried into practice. Perhaps, were a pair of fowls given a range of ten, twenty or more acres, and left to forage for themselves, one or two hens would be all the cock could attend; but confined to an acre or less, and fed on stimulating food, the bird's nature becomes, as it were, changed, and he feels himself qualified for greater deeds. We have seen a hen's back and sides all cut open by the cock's spurs, and the owner was complaining that the hen did not lay. If he had given her three or four companions his cause of complaint would have ceased. The only mode of deciding the question is by watching the fowls. We have known instances, though rare, of a cock serving twenty to twenty-five hens, and the eggs being very fertile. Again, a cock was cooped up with four hens, and it was found that when penned with two the eggs hatched twenty-five per cent. better than with the four. We think the latter case is of rare occurrence; a safe average is four to six hens to a cock. A few days' observation will enable one to tell whether more or less hens are needed. A young cock that has had a dozen or twenty hens to run with the first year is rarely fit for more than three or four the second. But if well cared for the first, and allowed not more than six hens, he is usually good for three or four years' service. We know many are prejudiced against using old cocks, and usually their prejudice is founded on experience like the above. A young cock with old hens is our preference for breeding stock, though many reverse it and put an old cock with young pullets. We know the hen lays a larger egg than the pullet, and a large egg must certainly bring out a larger chick than a small one; and, as a rule, (to which there are many exceptions,) a young cock is more vigorous than an old one. Therefore we think this selection preferable. Some, we are aware, contend that the cock has more influence on the progeny than the hen, and that an old cock, being more mature and developed, will throw better chicks. Such has not been our experience, however, after a close observation of several years' duration."

PREMIUM BIRDS DO NOT PRODUCE THE BEST CHICKENS.

For the purpose of more fully carrying out our idea of breeding fowls to perfection and pointing out their imperfections, we have selected the Brahma as an example, (the principle will apply to any other breed,) and in this connection give, from *Moore's Rural New-Yorker*, the experience and advice of a gentleman who makes the breeding of fowls a science. He says:—" Premium birds do not always produce the best chickens. Good results may often be obtained from moderate stock, provided that they be so selected that the defects of the cock may be counteracted by those of the hens. Size in the Brahma is not of so much importance as most people give to it. Fine, large chickens may be reared from small parents by proper care and attention, and good, regular and judicious feeding.

INFLUENCE UPON THE FANCY POINTS.

"The cock has the most influence upon the fancy points, while the hen has most upon the form and size. If more attention were paid to the shape and straightness of the comb of the cock, we should see less of those *grave* defects which so frequently mar whole pens. I have seen magnificent birds with such crooked and fungus-like combs as would almost disqualify them in my opinion. Judges have been too liberal with these defects. It is quite time such liberality was stopped. Crooked combs should be bred out and not tolerated. The comb is one of the most prominent characteristics of the bird, and almost the first object which meets the eye. It touches our sense of the beautiful immediately to see a small head and straight comb, and docile look. And the head of a Brahma fowl should possess these qualifications; too much importance should not be given to breeding for weight or largeness of carcass, over other qualifications. I admire in the Brahma fowl a large frame, of symmetrical proportions and corresponding weight; but a fattened fowl is only fit for the table. I should rather breed from a small cock with a perfect comb than a large one with a crooked comb. A lively cock, mated with large hens, is preferable to a sluggish cock and small hens. Length of legs in a cock is of less importance than in a hen; and in order to get size and proportion you must have *due* length of legs; and even in a hen, it may be counteracted by judicious mating. A narrow cock and a very wide hen are more likely to breed well than the reverse. It is to the male bird the breeder must look for perfection or defects in the comb, the beautiful yellow color of the legs, and all the fine points of the Brahma.

"As to the penciling, I am convinced, by considerable experience, that the two sexes bear a proportionate influence to each other, although I should not hesitate to say that there is more probability of breeding good chickens from a perfectly and darkly penciled pullet or hen and an inferior cock than from a badly colored or marked hen and a cock of superior blood. A hen with a bad comb, mated with a cock whose comb is small and fine, will throw some very fine chickens. A cock with a drooping back and saddle should be mated with a hen very high towards the tail; and if his hackle be short or scanty, that of the hen should be unusually sweeping and full. If any white stain should appear in the ear lobes, it is very apt to perpetuate itself, and particular care should be taken that the other sex has no sign of it, through several degrees. In shape, style and carriage, the Dark and Light varieties of the Brahma fowl should be precisely similar. In the Light, I think the breeders of this country have surpassed the English. The Light now stands almost on equality with the Dark in size, shape, and in general popularity. The comb of this fowl especially must be more closely looked after. A defective comb tells wofully against the bird. You must breed them even, low and straight. You cannot, I *know*, get this point to perfection in the cock until a strain has been bred for years. No *pure* strain ought to breed a comb in which the peculiar triple character is not perfectly distinct.

SHAPE OF THE COMB AND HEAD.

"There is a diversity of opinion as to the shape of the comb. It should not exceed half an inch in hight, and instead of rising from the front towards the back and ending in a peak, I should prefer to see it, after arising for half or two-thirds of its length, decrease again towards the back, forming a kind of arch. This kind of comb not only looks well and symmetrical, but according to experience, is likely to breed far more true than any other. The head of the Brahma cannot be too small in proportion to the body. There is no point in this fowl that so truly indicates the high breeding or the *blood* of the strain as the smallness of the head, and you will find that a small head is accompanied by fineness of flesh, a point never to be lost sight of in this class. I placed a dark hen of this variety in a coop by itself on exhibition at our poultry show merely to give those interested in the matter those points in perfection which I claim we must reach before we can say we have finished our labors in this respect.

THE GENERAL CHARACTERISTICS.

"In all the original Brahmas the *deaf-ears* fell below the wattles; and this point was mentioned by Dr. BENNET as a characteristic of the breed; and the perpetuation of this should be carefully looked after. The neck-hackle should start well out just below the head, making a full sweep, and marking the point of juncture between the head and neck very distinctly by an apparent hollow or depression. The hackles can hardly be too full, and should descend low enough to flow over the back and shoulders. The more perfect you can get this, the nobler the carriage and appearance of the bird. A short or scanty hackle is a very great blemish. The hocks should be well covered with soft curling feathers. A cock with hocks a little *out* should not be deprecated, and as sometimes is, by the inexperienced, discarded. This class of hock, when properly mated with fine built hens, scantily feathered on the legs and toes, throw very fine full-booted birds. While I should condemn all vulture-hocked fowls to the gridiron, there are exceptions where I have bred from a very large, finely-formed hen, with handsomely and distinctly marked pencilings, with great success, by mating them with a clean shanked cock with the proper marking; and have thrown four good birds to one hocked. No bird of this species should, when full grown, be considered fit for exhibition, unless the cock weighs twelve pounds, and hens from eight to nine pounds; and if a cockerel does not weigh eight pounds at six or eight months, he will rarely prove a show bird.

BREEDING AND MATING FOR SIZE, ETC.

"In breeding for size, select a short, compact, deep-bodied cockerel, which need not be large, and mate him with long backed hens, even if their legs are longer than usual. Although length of back is a decided fault, such a cross will generally breed well; the hen supplying the form, while the cock fills out to the proper proportion. Long, dangy, large-boned cocks may be mated with compact, short-legged hens, with the same result; but the first mentioned cross will produce better results. Fine chickens may be reared from the eggs of pullets; but the best chickens, as a rule, are got by mating either a two-year-old cock or a cockerel, with hens in their second season; their chickens fledge more quickly, and attain maturity sooner. Hens mated with cockerels turn out more male birds, while cocks mated with pullets, will produce a goodly proportion of pullets. I should not hesitate mating cockerels with pullets, if they be fine, strong-boned birds, hatched in March or the early part of April. A great many birds are spoiled by breeding from a cock of one strain and hens of different strains, and different styles of pencilings. If my presumption may be excused, I should advise the different breeders of this country to make up their minds respectively, as to the style and markings of the birds they deem most desirable to breed, and breed them uniformly and closely to the standard they have adopted. The popular taste will soon settle the question. You can always have fresh blood, if you keep two or three pens, and you can go on for years without crossing your breeds, and running the risk of bad blood or a motley brood, with no uniformity of shape or markings.

BREEDING IN-AND-IN.

"Do not feel too much anxiety about breeding *in-and-in*. Parent and offspring, and even brother and sister, may be bred from with safety and success for several years with this class of fowls. No breed has such stamina as the Brahma, and if any mishap does occur, it will not be so aggravated as it would be by the concentration of bad blood; therefore, it stands you in hand to be very careful what strain you purchase, and to know if the party has bred from distinct strains or indiscriminately. It is a work of time to breed fine strains, and considerable patience is requisite. It is in this respect that parties make a great mistake in going about from yard to yard, selecting here and there a bird from one, and cock, &c., from another, to gratify their vanity, with the hope of winning a few prizes, to the great detriment of the stock and disappointment of purchasers of the same, if they should breed from them. In the Light Brahma it is very necessary to secure a sufficient amount of color in the cock. The tendency of all poultry is to get lighter if indiscriminately bred; therefore, you should select cocks of the proper darkness for breeding stock. The saddle should only be lightly striped, for if it contains too much black or the neck-hackle too dark, you will produce spotted backs. I will set down two rules, either of which can be applied to suit the wants of the breeder:—1. Very heavy penciled cocks must be used to get heavy penciled (chicks) cocks. 2. Very dark hackled hens and light penciled hackled cocks will get nice hackled pullets."

VULTURE HOCKED FOWLS.

Vulture hocked birds are a disqualification to any brood of fowls, with few exceptions, and should be eschewed in all breeding stock. The vulture hock is the projection of feathers behind the knee, and inclining towards the ground, as shown in the accompanying illustration. The feathers of a fowl's leg usually should be close round the knee, and the leg clean below it. The breeds in which the vulture hock is necessary are Scrai-ta-ooks, Booted Bantams, and Ptarmigan fowls. Where the vulture hock makes its appearance, unwished  for, and where its presence is considered a grave fault, is among Cochins and Brahmas. The fault will sometimes appear in the progeny, but in fowls, as in everything else, the perfect birds form the exception, and as Dr. BENNETT says, " to have many of them it is only necessary to breed well and kill well. By this process you will get rid of the vulture hock."

CROSSING THE BREED.

As we have said elsewhere, to insure successful and beneficial crossing of distinct breeds, in order to produce a new and what may be considered a valuable variety, the breeder should be well versed in the laws of procreation, and the varied influences of parents upon their offspring. It is avered that all fowls bred in this country are crosses or made breeds, either by design or accident. Therefore crossing does not necessarily produce a breed; but on the other hand, it always produces a variety, and that variety becomes a distinctive breed only where there is a sufficiency of stamina to make a distinctive race, and continue a progeny with the uniform or leading characteristics of its progenitors. In crossing one breed with another we should say put a light cock with dark hens or *vice versa*, as in this case there is more liability of producing not only a new variety, but also some fine birds in the brood. Care is required in this matter, as in all others, (in mating for cross-breeding,) and patience is indispensable to success. All disqualified birds should be taken from the pen at the earliest moment, and sent to the table, leaving the best selections to breed from. We have made a fine cross by placing a dark Brahma hen with a white Dorking cock, and, on another occasion, made a good cross by placing a White-faced Black Spanish cock with a white Dorking pullet. There is no question but that good and valuable breeds of fowls, of beautiful plumage, may be thrown by these crosses.

SETTING HENS AND INCUBATION.

THE NUMBER OF EGGS TO PUT UNDER A HEN.

One of the most important points to be observed in setting eggs for hatching, is to correctly proportion the number, taking into consideration their size, and the size of the hen about to sit upon them. The state of the weather should also be a guide; for a hen capable of setting upon and hatching thirteen eggs in June ought not to have more than ten in January. The great error of setting a hen upon more eggs than she can cover is a cause of very general disappointment. We have frequently seen cross-bred game and other small hens set upon thirteen eggs, when it was perfectly clear to us that it would be impossible for them all to receive a proper and equal share of heat from her body. It is absolutely certain, also, that a hen cannot hatch out chickens from those eggs which she cannot draw close up to her body and give to them the natural warmth they require in the process of incubation. This has been very clearly demonstrated to us; for upon one occasion we placed fifteen eggs under a hen, when we ought not at any season to have given more than twelve, or, at the most, thirteen, and while out at feeding time, we examined the nest and found only thirteen eggs left. We at first thought the hen might have eaten them; but, after one or two examinations, we found sometimes thirteen and at others fourteen eggs present. We determined upon catching the hen one morning while off to feed, after finding there were only thirteen eggs in the nest. We cautiously laid hold of her, when she unfortunately dropped one egg and broke it; upon a further examination we found the other missing egg under her wing. We replaced the egg in the nest and found that she regularly removed one or two of them; thus it was apparent that she had more eggs under her than the surface of her body could possibly cover by contact. This marvelous fact proved the existence, first, of the beautiful principle we term instinct, and the ardent natural desire for carrying out to the fullest extent the remarkable operation we understand as incubation.

THE PROPER HENS TO SET.

A half-breed game or other small hen should be chosen for a natural incubator — (they have always, with us, proved the best breed) — and nine of her own eggs should be the extent; if a Dorking or a large size mongrel hen be selected, eleven are sufficient; a Cochin hen of some of the strains we have seen, will even cover fifteen of her own or eggs of similar size; but even in this instance, it is best to err on the safe side, and give her but thirteen eggs. Cochins and Brahmas have a large width of breast and a large amount of fluff and feather, both features being highly conducive to successful hatching, by assisting to retain the heat of the body of the birds and of the eggs also.

CLOSE-SETTING HENS.

There are some hens over-anxious about the chicks within the shells, whose cry for deliverance they can distinctly hear; and they do not rise from off the eggs during the process of chipping. This is an operation we have continually observed with hens that are very successful in hatching, while those which sit too closely at the last stages are those whose excess of kindness has produced the non, or limited, success in hatching out good broods. The only good arising from any sprinkling of the eggs with water results from their having received an increased and life-saving supply of air during such process, without which, in many instances, the chicks would either have been suffocated or glued to the shell.

THE PROCESS OF INCUBATION

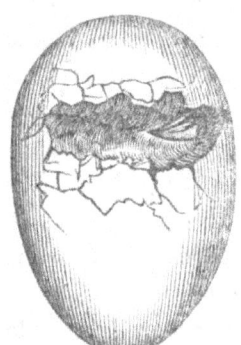

of the chicken is a subject not only curious but very interesting to the student of nature. It generally takes twenty-one days to hatch a brood of chickens, although a close-setting hen will sometimes hatch in eighteen days, if the weather is favorable. The expiration of the time should be carefully watched for; not that the chicken requires any assistance, but, on the contrary, interference is much more likely to prove an injury than a benefit. A healthy chick will perform all that is required to free it from the shell. It is wonderful the power they possess while rolled up in so apparently helpless a mass; the head, however, that makes the most exertion to free itself, is placed so as to leave room for reaction, and to turn round, and thus to peck a circle, (as shown in the accompanying engraving,) and breaks around the large end of the shell, admitting the air by degrees, until it becomes gradually prepared to extricate itself. A rash attempt to help them by breaking the shell, more particularly in a downward direction, toward the smaller end, is frequently followed by a loss of blood, which can ill be spared, and death ensues.

We place the nest in a warm, sheltered place, and have fresh food and water near at hand so that the hen can help herself whenever she is so inclined. Should the nest become dirty, change it, or even wash the eggs in tepid water. As fast as the chickens break the shell, place them in a basket of cotton-wool by the fire, to avoid the danger of the mother's crushing them while they are helpless. When all have hatched, they may be returned to the hen. The yolk of a hard boiled egg should constitute their food dur ing the first week; after which coarser food may be given. When fully fledged, give them their liberty in the heat of the day, and house them before sunset. Never permit them to wander in the grass when the dew is on, as more healthy fowls perish from this than any other cause. The chicks can be fed to good advantage with cracked corn or a mush of potatoes and Indian meal cooked. Feed should be given in small quantities, and frequently, during the day.

CHANGES WHICH AN EGG UNDERGOES IN HATCHING.

In this connection we trust it will not be deemed out of place to give what we find in an old volume of the *Genesee Farmer and Gardeners' Journal* of July, 1833, relative to the wonderful changes which an egg undergoes in hatching, from the first day till its final exclusion, accompanied with three illustrations, showing the first, middle and last stages of the chick. The same article appears in the *American Poulterer's Companion*, erroneously credited to an English journal. This process of incubation is thus minutely described:

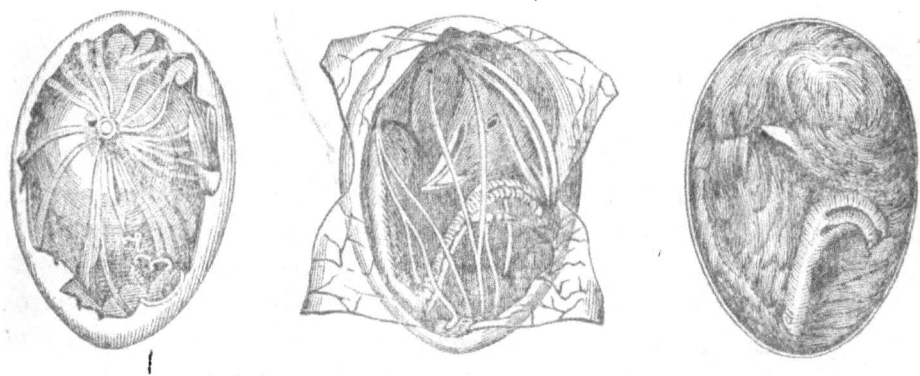

FIRST, MIDDLE, AND LAST STAGES OF THE CHICK.

"The hen has scarcely sat on her eggs twelve hours before some lineaments of the head and body of the chicken appear. The heart may be seen to beat at the end of the second day; it has at that time somewhat the form of a horseshoe, but no blood yet appears. At the end of two days, two vesicles of blood are to be distinguished, the pulsation of which is very visible; one of these is the left ventricle, and the other the root of the great artery. At the fiftieth hour, one auricle of the heart appears, resembling a noose folded down upon itself. The beating of the heart is first observed in the auricle, and afterward in the ventricle. At the end of seventy hours, the wings are distinguishable; and on the head two bubbles are seen for the brain, one for the bill, and two for the fore and hind part of the head. Toward the end of the fourth day, the two auricles already visible draw nearer to the heart than before. The liver appears toward the fifth day. At the end of a hundred and thirty-one hours, the first voluntary motion is observed. At the end of seven hours more, the lungs and the stomach become visible;

and four hours afterward, the intestines, and loins, and the upper jaw. At the hundred and forty-fourth hour, two ventricles are visible, and two drops of blood instead of the single one which was seen before. The seventh day, the brain begins to have some consistency. At the hundred and nineteenth hour of incubation, the bill opens, and the flesh appears in the breast. In four hours more, the breast-bone is seen. In six hours after this, the ribs appear, forming from the back, and the bill is very visible, as well as the gall-bladder. The bill becomes green at the end of two hundred and thirty-six hours; and if the chicken be taken out of its covering, it evidently moves itself. The feathers begin to shoot out toward the two hundred and fortieth hour, and the skull becomes gristly. At the two hundred and sixty-fourth hour, the eyes appear. At the two hundred and eighty-eighth, the ribs are perfect. At the three hundred and thirty-first, the spleen draws near the stomach, and the lungs to the chest. At the end of three hundred and fifty-five hours, the bill frequently opens and shuts; and at the end of the eighteenth day, the first cry of the chicken is heard. It afterward gets more strength and grows continually, till at length it is enabled to set itself free from its confinement.

"In the whole of this process we must remark that every part appears at its proper time; if, for example, the liver is formed on the fifth day, it is founded on the preceding situation of the chicken, and on the changes that were to follow. No part of the body could possibly appear either sooner or later without the whole embryo suffering; and each of the limbs becomes visible at the first moment. This ordination, so wise and so invariable, is manifestly the work of a Supreme Being; but we must still more sensibly acknowledge His creative powers, when we consider the manner in which the chicken is formed out of the parts which compose the egg. How astonishing it must appear to an observing mind, that in this substance there should at all be the vital principle of an animated being; that all the parts of an animal's body should be concealed in it, and require nothing but heat to unfold and quicken them; that the whole formation of the chicken should be so constant and regular that, exactly at the same time, the same changes will take place in the generality of eggs; that the chicken, the moment it is hatched, is heavier than the egg was before! But even these are not all the wonders in the formation of the bird from the egg — for this instance will serve to illustrate

the whole of the feathered tribe — there are others altogether hidden from our observation, and of which, from our very limited faculties, we must ever remain ignorant."

THE FERTILITY OF EGGS.

There is no difficulty whatever in testing the fertility of eggs. The way to ascertain unfertile eggs at as early a period as possible is to take them into a room moderately dark, and hold them between the eye and a candle or lamp, in the manner represented in the engraving on the preceding page. The eggs under a setting hen should be examined at least as early as the eighth day after she commences incubation. If the egg be fertile, it will appear opaque, or dark all over, except, perhaps, a small portion towards the top; but if it be unimpregnated, it will be still translucent, the light passing through it almost as if new laid. After some experience the eggs can be distinguished at an earlier period, and a practiced hand can tell the unfertile eggs even at the fourth day. Should the number withdrawn be considerable, four batches set the same day may be given to three hens, or even two, and the remainder given fresh eggs; and if not, the fertile eggs will get more heat, and the brood come out all the stronger.

THE PROPER FOOD AND FEEDING.

NEVER stint poultry in the variety or quality of their food. Good food is positive economy. The best and heaviest corn is the cheapest. The best food is that which gives the most of what nature demands for the formation of muscle, bone and fat. Fine bran, or middlings, is richer in two of these important ingredients than any other one kind of food; but being deficient in gluten, is not warmth-giving, and is better when combined with whole grain, which, when mashed, forms a most wholesome and nutritious diet. Barley is much used in Europe, but should never be the only food in the poultry yard. Fowls do not fatten on it, though for a time they will thrive. Oats are good as a change, but inferior in nutriment; if they are browned or roasted and given freely, they prove a good egg-producing food. Buckwheat, however, is the best food to make fowls lay early. They devour the feed greedily, and its heating influence, in winter, is very perceptible. Hemp seed is also productive of eggs, and is very strengthening; it is one of the best things that can be fed to fowls during the moulting season.

THE PROPER FOOD TO GIVE.

In preparing birds for exhibition, flax seed may be given occasionally; it increases the secretion of oil, and gives luster to their plumage. In giving soft feed it should be mixed stiff — not mushy; fowls do not relish it in the latter state. A good food of this kind is composed of equal parts of fine bran and Indian meal. This should be scalded or mixed with boiling hot water to such a consistency that it will break or crumble when thrown upon the ground. Another good soft feed is made of small potatoes, washed clean, boiled, and mashed with an equal quantity of Indian meal. In giving soft feed never use a feeding dish or trough. If the yards are clean, as they should be, the ground is by far the best place to feed them from. The gravel and sand, which adhere to the food, are necessary for digestion; besides, poultry prefer to pick their food from the ground.

Do not, on any consideration, neglect to give poultry green food. A little chopped vegetables of some kind, whether cabbage, lettuce, spinach, onions or other greens, is better given every day than a great deal once or twice a week. To secure perfect eggs, lime, in some form, ought to be furnished. Broken bones, lime rubbish, oyster or clam shells, burned and pounded fine, are all good. Beef or pork scraps are productive of good results. In the winter, when fowls cannot supply themselves with insects, worms or grubs, a scrap-cake, laid in the hen yard for them to pick at, or a little chopped off and broken up and fed to them, adds not only to their health but largely to the contents of the egg basket. An occasional dish of raw meat, chopped into small pieces and given them will be devoured with avidity. Another way, and one which we have practiced with good results, is to get a sheep's pluck and hang it up in the hennery, just high enough to make the fowls fly up and pick it off by piece-meal. If fowls are over-fed with meat it will show itself in the loss of feathers, and prove very detrimental to the brood. Some breeders feed game fowls largely on fresh meat — claiming that it creates a pugnacious disposition in the cock. Whatever is done in the matter of feeding, regularity, as to time, is essential to success.

REARING FOWLS FOR MARKET AND EGGS.

THE BEST BREED TO REAR FOR MARKET.

The best breed of fowls to rear for the market, or as egg-producers, depends upon locality; for while, in some places, one variety is deemed the best, in others it would prove the reverse. Our own opinion is, that, for a market fowl, the Brahmas and Cochins will, under almost all circumstances, prove the most desirable, they being less liable to disease, feathering up quickly, and can be bred to weigh, at from four to six months of age, eight to ten pounds. Another good table fowl is the Dorking (cock) crossed with the Brahma (hen). The flesh of this cross is sweet and nutritious, and acquires at early age the plumpness of the Dorking at maturity. There are other breeds, however, which are said to be desirable to rear for the table. Many claim that the French breeds of fowls are of this number; but this we very much doubt, as their flesh lacks the buttery, golden color that attracts the eye of the epicure. They may prove valuable as egg-producers, but they lack many good qualities as a table bird. Dorkings are undoubtedly at the head of the list as table birds, but of late years have become so subject to disease that we question the feasibility of rearing them profitably for market in our changeable northern climate.

THE BEST AS EGG-PRODUCERS.

As egg-producers the Hamburgs are claimed to stand at the head of the list. This claim we are prepared to dispute; for, as winter layers, we find that the Brahma, Cochin, Leghorn, Poland, and Houdan stand relatively in the position here named. That the Hamburgs are good egg-producers we admit; but that they are any better than a number of non-setting fowls, so called, we deny. The richness and meatiness of their eggs are not to be compared with those of the Poland, Leghorn, Houdan or Brahma; and their eggs lack the size of those named. All things considered, we have no hesitancy in saying that for eggs we should name the Polands; for the table, Dorkings, and for early marketable chickens, Brahmas and Cochins.

A correspondent of *Moore's Rural New-Yorker*, who has had considerable experience in rearing fowls for profit, says:—"The Farmer's Breed is the breed for profit. It consists of Brahma hens and colored Dorking cocks — the chicks from which are hardy, easily reared, grow fast, and in four months, without extra feed, will dress four to five pounds each of fine-grained, well-formed, plump-breasted, well-colored flesh, fit for the table of any amateur or epicure, and always commanding a good price in market. The hens from this cross are even better and more continuous layers than either pure Brahma or the Dorking; but if wanted to breed again, the farmer must keep one coop separate of Brahmas — say a cock and two hens — and so also of the Dorkings, and thus yearly with the cross of pure bred birds, cocks of the Dorkings, and hens of the Brahmas, keep up the '*Farmer's Breed for profit.*'"

FATTENING AND PREPARING POULTRY FOR MARKET.

THE MANNER OF FATTENING.

ALTHOUGH the manner of fattening poultry may seem to be extremely plain, there is, nevertheless, a right and a wrong way, a long and a short mode of accomplishing the object desired. Many breeders who rear fowls for the market believe in letting poultry forage and shift for themselves, while others believe the best method is in keeping them constantly in high feed. This is just our idea; for where a steady and regular profit is required from rearing poultry, or a business is made thereof, the very best method, whether for domestic use or for the market, is constant high keep from the beginning. Thus they will always be in a saleable condition and ready for the table. As the *American Poulterer's Companion* justly says, fowls kept in this way need but very little extra attention. Their flesh will be superior in juiciness and richer in flavor than those which are fattened from a low and emaciated state. Fed in the manner above indicated, spring pullets are particularly fine, commanding the highest price on the market, and proving a most healthful, nourishing and restorative food.

FEEDING HOUSES.

Our mode of constructing feeding houses or coops is to have them so they will be at once warm and airy, with earthen floors, well raised, and capacious enough for the accommodation of from twenty to thirty-five fowls; the floor, if desired, may be slightly littered with straw, but the litter should be frequently changed, and great care taken to secure cleanliness, for fear of vermin. As we have before said, the coops should be well supplied with feeding-troughs which should always be kept *full of feed*, and which can be got at easily by the fowls. Perches should also be placed but a few feet from the ground, so they can be reached without much effort; those made in the form of stairs, having the poles one above the other, (slanting,) are the best. Fowls cooped in this way may be fattened in a short time and to the highest pitch, and be preserved in a perfectly healthy state. There is no necessity, in our opinion, to confine fowls in dark coops and practice the art of cramming to fatten them properly; this mode is an abomination, and should not be followed by any breeder of common sense.

MODE OF FATTENING FOWLS IN COOPS.

In fattening fowls confined in coops, old writers recommend feeding them with bread, soaked in ale, wine, or milk; barley mixed with milk, and seasoned with mustard or anise seed; while others recommend cramming them three or four times a day; also keeping them in a dark place, and not allowing them any exercise. BRADLEY says, "the best way, and the quickest, to fatten them, is to put them into coops as usual, and feed them with barley meal, being particular to put a small quantity of brick dust in their water, which they should never be without. This last will give them an appetite for their meat, and fatten them very soon." Yet another writer says they should be shut up where they can get no gravel; keep corn by them all the time, and also give them dough enough for one feed a day. For drink, give them skimmed milk; with this feed they will fatten in ten days; if they are kept over ten days, they should have some gravel, or they will fall away.

The mode of fattening poultry, extensively practiced in Liverpool, England, is to feed them with steamed or baked potatoes, *warm*, three or four times a day; the fowls are taken in good condition from the yard, confined in dry, well-ventilated coops, and covered in, so as to prevent the entrance of too much light. It is said this method is attended with the greatest success.

NO POULTRY SHOULD BE PERMITTED TO RUN AT LARGE

for at least ten days before killing, for they are apt to range in the barnyards, and pick up filthy food, which permeates all through the bird, and frequently they become so tainted that they are unfit to eat, after being placed on the table.

PROPER FOOD FOR FATTENING.

In all cases in fattening fowls, whether old or young, we should recommend that the food be cooked and fed *warm*. Barley meal, or mixed with equal quantities of Indian meal, made into a thick paste or porridge and fed warm, is about as good a feed as we know of, and seems to make flesh faster and more solid, and give it a golden color and plump appearance after being dressed.

KILLING AND DRESSING.

As much if not more depends on the manner of killing poultry as in the dressing to have it look fit for market. Too much caution cannot be used in this branch of the business. One mode of killing fowls, (instead of wringing the necks, which we deprecate,) is to cut their heads off with a single blow of a sharp ax, hang them up by the legs, and allow them to bleed freely, and pluck their feathers immediately — while *warm*. The French mode, which is highly commended, we think far the best, as it causes instant death without pain or disfigurement, and is simply done by opening the beak of the fowl, and with a sharp-pointed and narrow-bladed knife, make an incision at the back of the roof, which will divide the vertebræ and cause immediate death, after which hang the fowl up by the legs till the bleeding ceases, and pick it while *warm*, if you desire the feathers to be removed. With a little care the skin of the fowl does not become as torn and ragged as it does in the old-fashioned way of scalding. Another thing, the flesh presents a better and more natural appearance when not scalded.

GEYELIN says:—"Some breeders cram their poultry before killing, to make them appear heavy; this is a most injudicious plan, as the undigested food soon enters into fermentation, and putrefaction takes place, as is evidenced by the quantity of greenish, putrid-looking fowls that are seen in the markets." Fowls should *always* be allowed to remain in their coops at least twenty-four hours previous to being killed, without food; by so doing, the breeder will be the gainer in the end, as his poultry will keep longer and present a better appearance in the market; and, above all, he will show the purchaser that he is honest, and has not crammed his poultry for the purpose of benefiting himself and swindling others.

THE FRENCH MODE OF KILLING

is preferable, when the head of the bird is to be left on; but that is not necessary, neither is it desirable; but when the head is taken off, the skin should always be pulled over the stump and tied. The mode of picking while the bird is warm is called "dry picking," and is the favorite method of dressing poultry for the Philadelphia market. There is one objection to this system, that it does not improve the appearance, although it does the flavor; and while cooking it will "plump up" and come out of the oven looking much finer than when it went in. In addition, it will keep much longer than when dressed by the other mode. Another plan is, after the bird is picked, as above described, plunge it in a kettle of very hot water, holding it there only long enough to cause the bird to "plump," then hang it up, turkeys and chickens by the foot, and geese and ducks by the head, until thoroughly cooled. This scalding makes the fat look bright and clear, and the fowl to appear much fatter than it would if picked dry. This is the usual mode of dressing for the New York markets.

BOXING POULTRY FOR MARKET.

On the subject of boxing poultry for market Dr. BENNET says:—"It should be carefully packed in baskets or boxes, and above all, it should be kept from the frost. A friend of mine, who was very nice in these matters, used to bring his turkeys to market in the finest order possible, and always obtained a ready sale and the highest market price. His method was to pick them dry, while warm, and dress them in the neatest manner; then take a long, deep, narrow, tight box, with a stick running from end to end of the box, and hang the turkeys by the legs over the stick, which prevents bruising or disfiguring them in the least." The way poultry is frequently forwarded to city markets is enough to disgust almost any one, and throws odium on breeders as a class.

THE MODE OF PACKING.

All poultry should be thoroughly cooled before packing. Then provide boxes, for they are preferable to barrels; place a layer of rye straw that has been thoroughly cleaned from dust, on the bottom. Commence packing

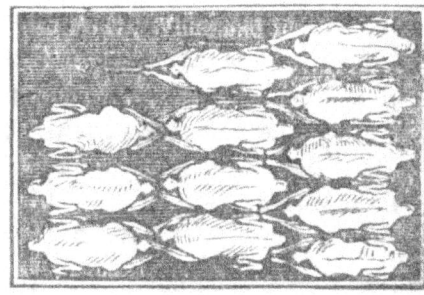

FIG. I.　　　　　　　　　　FIG. II.

by bending the head of the fowl under it (see figure 1.) Then lay it in the left hand corner, with the head against the end of the box, with the back up; continue to fill that row in the same manner until completed; then begin the second row the same way, letting the head of the bird pass up between the rump of the two adjoining ones, which will make it complete and solid, (see figure 2.) In packing the last row, reverse the order, placing the head against the end of the box, letting the feet pass under each other; should there be a space left between these two rows wide enough to lay in a few sideways, do so, passing the feet under the same way, but should it not be wide enough, then fill tight with straw, so the poultry cannot move. This gives a uniformity of appearance, and a firmness in packing that will prevent moving during transportation. Over this layer, place straw enough to prevent one layer from coming in contact with the other; then add other layers, packed in the same manner, until the box is filled.

Care should be taken to have the box filled full, in order to prevent any disarrangement of the contents; for should they become misplaced, the skin may become so badly disfigured as to cause a depreciation of the value to the owner. Great care should be taken in packing not to skin the bird.

for during transportation, the skinned places turn black and make it look badly. To those having extra fine poultry to send to market, we would recommend to put paper over each layer before placing the straw on it; this prevents the dust settling on it, and adds much to its appearance. A little practice will soon make a person quite expert in packing, and for a person buying to ship an expert packer is valuable; his skill will pay the owner ten times his cost, for very frequently the first sight of a box of poultry sells it.

MARKING THE BOXES, ETC.

The box should have the initials of the consignor, the number and variety of the contents, as well as the name of the consignee, marked on it. The necessity for marking the number and variety of contents is, that in case the box is broken open and any portion of the contents missing before delivery to the consignee, they will be enabled to make a correct bill for the missing poultry. Another advantage is, that the consignee knows by a glance at the box whether it contains the desired variety he wishes; if not, he need not open it, and the contents will not receive a needless handling; for some parties prefer a mixed box, while others do not, and all dealers prefer selling the entire contents of the box to one person, as it avoids error in weighing and keeping the accounts. To those wishing to market capons we would say, they should be dry picked, with the feathers on around the head and the tip of the wings; also the tail feathers left in; the small or pin feathers should all be removed.

SEND GEESE FOR CHRISTMAS,

as they are in demand at that time, and bring more money than any other poultry. All Irishmen and many Germans think it is not Christmas without a goose for dinner. Send all large turkeys before New Year's, as they are wanted to adorn the New Year's table; and they depreciate in price immediately after that day. Small turkeys are then in better demand, while chickens and ducks can be sent any time after they are fattened, and never until then.

Persons living at a distance from the city and desiring to send their poultry to market for any particular occasion, should allow at least two days longer for its transportation than usual, so that it will not miss the market for that occasion; for the dealer had better receive it a day or two sooner than one hour too late.

PURCHASING POULTRY FOR THE TABLE.

As we have given the *modus operandi* for fattening fowls for market, &c., we now have a word of caution to offer those purchasing poultry which may not, perhaps, come amiss. Those who are not good judges of poultry, as to their age, may, and often do, have old, tough fowls palmed off upon them by an unscrupulous dealer.

Fowls are killed and prepared for market with much adroitness and care by some dealers, and many devices practiced to catch the eye of the unsophisticated purchaser — the best side of the poultry being shown to the greatest advantage. Every sort of fowl is killed, plucked and put on the market, and if the purchaser buys an inferior article at an exorbitant price, he has only himself to blame for so doing.

HOW TO JUDGE THE AGE OF POULTRY.

The age of a plucked fowl can be judged simply by the legs. If the scales on the leg of a hen are rough and the spur hard, it will not be necessary to see the head to determine that she is *old ;* still the head will corroborate your observation; if that of an old hen, the bill will be stiff and hard, and the comb rough and thick. The scales on the leg of a young hen are smooth, glossy and fresh colored, whatever the color may be; only the rudiments of spurs are observable; the claws tender and short, the under bill soft, the comb thin and smooth. An old hen turkey has rough scales on the legs, callosities on the soles or bottom of the feet, and long, strong claws; while a young turkey has the reverse of these marks. A young goose or duck can be readily told by the tenderness of the skin under the wings, the strength of the joints of the legs, and the coarseness of the skin.

If the foregoing directions are strictly followed, in purchasing poultry, we will venture the assertion that the "good housewife" will have no fault to find with the length of time it takes to cook, or the toughness of her Thanksgiving turkey, goose or chicken. This mode of finding out the age of fowls is infallible.

PURCHASING UNDRAWN POULTRY.

We are one of a score of housekeepers who object, in toto, to the purchasing of poultry unless it be drawn. The habit of forcing fowls on the market undrawn, and allowing them to freeze and thaw, (generally with full crops,) by which process they become fetid and turn green cannot prove otherwise than unwholesome food — not fit to be eaten. No fowls should be purchased by housekeepers unless they are properly cleaned and drawn. In many cities there is a fine imposed upon the person for offering undrawn poultry upon the market for sale.

TO PRESERVE POULTRY IN WINTER.

This is a matter not fully understood, and for the information of the general reader we cannot do better than to give the mode practiced by the venerable Judge BUEL, in preserving poultry in winter. He says:—" I purchased a quantity of poultry for winter use early in November. The insides were carefully drawn, their place partially filled with charcoal, and the poultry hung in an airy loft. It was used through the winter, till about the first of February, and although some were kept seventy days none of it was the least affected with must or taint, the charcoal having kept it perfectly sweet."

VARIETIES OF FOWLS.

HISTORY, CHARACTERISTICS, Etc., OF THE BREEDS.

THE BRAHMAS.

WE have seen Brahmas which we considered the *ne plus ultra* of the feathered tribe. English breeders claim everything that is good for these

DARK BRAHMA COCK.

birds, and lose sight of their faults. We have bred the Brahmas, both Light and Dark, and thought highly of them; still they did not prove good layers with us. Since we have discarded them we have found out the reason of our ill-success — it was over-feeding. This may seem strange, but nevertheless it is a fact. We fed them all they could eat "and more too." The consequence was we did not get from them the number of eggs we otherwise

DARK BRAHMA HEN.

should. Feeding fowls enough to keep them in good heart and over-feeding them are two different things. In the first instance you are "just and generous" with them, in feeding just enough — in the other case you are "killing them with kindness" by over-feeding, which makes them dumpish and lazy, and inclined to be perpetual sitters. We believe that Brahmas well kept will make a very profitable fowl to breed. They are good layers, good sitters, and make the best of mothers, if the breeder knows how to handle them. They are objected to by many poultry fanciers, from their clumsiness — many aver that they are liable to break their eggs, when sitting, by getting off and on their nests. If the nests were put in the proper place, this fault would be obviated. Always make the nests low — on the ground or floor of the hennery is best; nail cleats around them of two-inch boards, not higher than two and one-half inches, to keep the eggs from rolling out, and you need have no fears of any being broken.

It is said the Brahmas are an Asiatic breed of fowls, and that they were first brought to this country by a sailor, who said he got them from the banks of the Brahmapootra — a river that waters the territory of Assam. How true this is we cannot say, but it is claimed that the Brahmas in this country sprung from this source, and that English breeders are indebted to America for the beautiful fowls of this breed they possess. These birds are highly prized in England — a pair of them having lately been sold for $350.

DARK BRAHMAS.—The Dark Brahmas are claimed by many breeders to be the best of the Brahma variety, but we opine there are just as many who stand ready to claim that the Light are equally as good, if not a better breed. Still some breeders claim that the flesh of the Dark is richer and more palatable than that of the Light. Our opinion is that the difference between the two colors is all *fancy*, one proving just as good as the other, under similar management. Having bred both colors, we have yet to learn the *distinctive* difference between them. The plumage of the Dark does not show the same mussiness of feather as the Light; still, if kept in a clean, dry hennery, as fowls always should be, the difference is imaginary.

The head of the cock should be surmounted with what is termed a "peacomb," which resembles three small combs running parallel the length of the head, the center one the highest; beak strong, well curved; wattles full; ear-lobes red, well rounded and falling below the wattles. The neck should be short, well curved; hackle full, silvery white striped with black, flowing well over the back and sides of the breast; feathers at the head should be white. Back very short, wide and flat, rising into a nice, soft, small tail, carried upright; back almost white; the saddle feathers white, striped with black, and the longer the better. The soft rise from the saddle to the tail, and the side feathers of the tail to be pure lustrous green black, (except a few next the saddle,) slightly ticked with white, the tail feathers pure black. The breast should be full and broad, and carried well forward; feathers black, tipped with white. Wings small, and well tucked up under the saddle-feathers and thigh fluff. A good black bar across the wing is important. The fluff on the hinder parts and thighs should be black or dark gray; lower part of the thighs covered with soft feathers, nearly black. The markings of the hen are nearly similar to those of the cock. Both sexes should have rather short yellow legs, (those of the hen the shorter,) and profusely feathered on the outside. The carriage of the hen is full, but not so upright as that of the cock. The markings of the hen, except the neck and tail, are the same all over, each feather having a dingy white ground, closely penciled with dark steel gray, nearly up to the throat on the breast.

LIGHT BRAHMA COCK.

LIGHT BRAHMAS.—Pure Light Brahma fowls are chiefly white in color of plumage, but if the feathers are parted, the bottom of the plumage will appear of a bluish-gray, showing an important distinction between them and White Cochins, in which the feathers are *always* white down to the skin. The neck-hackles should be distinctly striped with black down the center of each feather. The plume of the cock is often lighter than that of the hen; the back should be quite white in both sexes. The wings should appear white when folded, but the flight feathers are black; the tail black in both cock and hen; in the cock, however, it is well developed, and the coverts show splendid green reflections in the light; it should stand tolerably upright, and open well out laterally, like a fan; the legs should be yellow and well covered with white feathers, which may or may not be very slightly mottled with black; ear-lobes must be pure red, and every bird should have a perfect pea-comb, though fine birds with a single comb have occasionally been shown with good success; but, as a general thing, the pea-comb fowl shows off to the best advantage, and attracts universal commendation by both the amateur and breeder.

THE CHITTAGONGS.

Years ago this breed of fowls was looked upon as possessing a great deal of merit, but in these latter days of Brahma and Cochin fever they have been lost sight of, and we scarcely hear the name of Chittagong mentioned; though we firmly believe the Buff and White Cochins owe their parentage to a cross with the Chittagong and Shanghae breed. KERR's "Ornamental Poultry Breeder" says the plumage of the Chittagong is very showy and of various colors; the birds being exceedingly hardy. In some, gray predominates, interspersed with lightish yellow and white feathers in the pullets; the legs being of a reddish flesh-color, and more or less feathered; the comb large and single; wattles very full, wings good size; the model is graceful, carriage proud and easy, and action prompt and determined. The flesh of this breed is delicately white. The cocks, at eight or nine months of age, weigh from nine to ten pounds, and the hens from eight to nine pounds. They do not lay as many eggs during the year as smaller hens, but they lay as many pounds as the best breeds. The Red variety of Chittagongs are smaller than the gray; legs being yellow and blue; the wings and tail short; comb single and rose-colored. An ordinary pair will weigh from sixteen to eighteen pounds. In the dark-red variety the cock is black on the breast and thighs; the hens yellow or brown, with single serrated comb; legs yellow and heavily booted with black feathers. The Chittagongs as a breed is quite leggy, in many instances, the cock standing twenty-six inches high, and the hens twenty-two.

THE COCHIN CHINAS.

Birds of this breed are becoming more and more favorites with the general breeder, not only in England, but also in this country. They are deservedly high in the standard of merit in this country on account of their hardiness and good laying and breeding qualities. A friend of ours who has had considerable experience with the Asiatic breeds of fowls, considers the Buff Cochins better adapted to our severe and changeable climate than either the Brahma, Chittagong or Shanghae. He avers that they (the Cochins)

BUFF COCHIN COCK.

require less care, and pay for their feed in extra amount of flesh, and richness, and quality of eggs. His hens have weighed ten pounds each, and the cock fifteen pounds, and stands over two feet in hight. He allows his hens to have but one good setting a year, and breaks up this propensity in about two or three days by removing them to a coop with a bottom made of rollers two inches in diameter, and gives them little or no feed and fresh water. He thinks a hard bed a good cure for indolent habits. Although called Cochin Chinas, the Buff Cochins are the real Shanghaes. They were unknown to the Southern Chinese, and they never claimed them as their native fowl, and were as much astonished at their size as we were when they first came to this country. The Shanghae breed had feathered and unfeathered legs, but were more frequently unfeathered. Fashion, however, calls for booted-legs. There are three varieties of color—Buff, Lemon and Cinnamon. The Buff

BUFF COCHIN HEN.

seem to be the most desired. There are also Silver Buffs and Silver Cinnamons. The latter, if well marked, are very beautiful and rare.

The carriage of the cock should be upright and majestic; breast very broad, forming a straight line from the crop to the thighs; back short and wide; tail very slightly raised, and the wings very short and held tightly to the sides; the legs, thighs and saddles unusually large in proportion to the rest of the body; head small and carried well up; a stout, curved and yellow beak, with plenty of substance at the base, and the shorter the better. The carriage of the hen must be similar in general character to the cock, excepting that the head is carried much lower; and a gentle, pleasing expression of face is a mark of high bred specimens. The hackle of the cock should be very full and of a light bay color, spreading over the base of the wings and free from any markings. The hen's hackle should be a distinct, clear buff, free from any markings; a slight penciling is preferable to a clouded one. The saddles of the cock and hen should also be free from any markings. Cockerels of the year, though imperfect, will, if of pure blood, in the second year moult out perfectly clear. A black tail in the cock is ad missible; but the principal feathers, if bronze in color, add very much to the

PAIR OF PARTRIDGE COCHINS.

appearance of the bird; if of buff color, will throw dark pullets. The breast of the cock and hen should be clear buff, the feathers running somewhat lighter in color towards the tip, showing a waving appearance in sunlight. Both *primary* and *secondary quills* should be clear buff, without admixture of colors. The legs should be very heavily feathered, short, and wide apart. The comb in cock and hen should be very flat, evenly serrated and perfectly straight, without any inclination to either side. The wattles of the cock thin and fine, perfectly florid in color, ear-lobes well developed, long, thin and fine; any white is a decided blemish. The eye of the cock should be yel *low-ochre* colored; in the hen a little darker than those of the cock; and, strange to say, these characteristics denote a sound constitution. A clear, dark-winged cock throws the best chickens. Vulture hocks in Cochins are clearly inadmissible, and should never be tolerated at any exhibition; they show mixed blood, and, if bred out, will revert back again. Hocked birds are frequently awarded the highest premium at shows in this country — in England they are disqualified.

THE SHANGHAES.

The Shanghae fowl was highly estimated on its first introduction in this country in 1847, and for a long time thereafter considered the best of the Asiatic breed, but of late years we hear very little mention made of them. They are entirely ignored even from our poultry shows. As we have said elsewhere, the Cochins have superseded the Shanghae breed entirely. A well-bred cock, when full-grown, stands twenty-eight inches high; the hen from twenty to twenty-three inches. The hen has a slightly curved beak, the forehead well arched; comb low, single, erect, slightly and evenly toothed; wattles small and curved inward, the eyes are bright and prominent, the neck about eight inches long and gently arched when held upright; the body long and greatly arched; the girth of the body of a good specimen, when measured over the wings, is about twenty inches; the legs are rather long, of a pale yellow color, with a tinge of flesh-color, and generally thickly covered with feathers from the outside down to the toe. The plumage is remarkably soft and silky, and, beneath the tail, densely fluffy and rounded. The comb of the cock is high, deeply indented, and his wattles double and large. Though the comb and wattles are not to be regarded as the chief characteristics of this breed of fowls, nor are its reddish-yellow feathered legs; but the abundant, soft and downy covering of the thighs, hips, and region of the vent, together with the remarkably short tail, are characteristics not found in any other bird. The wings are small and short in proportion to the size of the fowl, being carried very high up the body, thus exposing the whole of the thighs, and a large portion of the side. The arrangement of the feathers gives the bird a greater depth of quarter, in proportion to the brisket, than any fowl with which we are conversant. There are Shanghae fowls of Black, Gray, Buff, Cinnamon and Partridge-color. These are termed sub-varieties. White is said to have been the color of the original imported birds, the other colors having been bred in this country. Mr. BOWMAN, an eminent English breeder of the Shanghae, says of the fecundity of this breed, that he had "a pullet that laid one hundred and twenty eggs in a hundred and twenty-five days, then stopped six days, then laid sixteen eggs more, stopped four days, and again continued her laying." The eggs are not so rich and nutritious as those of the Dorking; neither are they remarkably large compared to the size of the fowl; they are of a pale yellow or nankeen color, and generally blunt at the ends. The flesh of the Shanghae is quite inferior to that of the smaller breeds, being coarse-grained, neither tender nor juicy, and have more offal and less breast-meat than either Cochins or Brahmas. They are not inclined to ramble, and, on this account, bear confinement much better than many other breeds.

The White Shanghae.—This variety is entirely white, with the legs usually feathered, and differs in no material respect from the red, yellow, and Partridge, except in color. The legs are yellowish, or reddish-yellow, and sometimes of flesh-color. Many prefer them to all others. It is claimed by the friends of this variety that they are larger and more quiet than other varieties, that their flesh is much superior, their eggs larger, and the hens more profitable. Being more quiet in their habits, and less inclined to ramble, the hens are invaluable as incubators and nurses; and the mildness of

PAIR OF WHITE SHANGHAES.

their disposition makes them excellent foster-mothers, as they never injure the chickens belonging to other hens. These fowls will rank among the largest coming from China, and are very thrifty in our climate. A cock of this variety attained a weight of eight pounds, at about the age of eight months, and the pullets of the same breed were proportionably large. They are broad on the back and breast, with a body well rounded up; the plumage white, with a downy softness — in the latter respect much like the feathering of the Bremen goose; the tail-feathers short and full; the head small, surmounted by a small, single, serrated comb; wattles long and wide, overlaying the cheek-piece, which is also large, and extends back on the neck; and the legs of a yellow hue, approaching a flesh-color, and feathered to the ends of the toes.

THE MALAYS.

This breed of fowls is very large and clumsy, and possesses no particular merits that we are aware of, unless it be in size. They are decidedly Shanghaeish in appearance and action. The usual hight of the cock is from twenty-six to twenty-eight inches, and weighs on an average from ten to twelve pounds. We reared the fowls in 1857 on a small scale, and found them in attitude uncouth, their gait being heavy and destitute of alertness. WRIGHT says of this breed, that "in form and make they are as different from Cochins as can well be. They are exceedingly long in the neck and

PAIR OF MALAYS.

legs, and the carriage is so upright that the back forms a steep incline. The wings are carried high, and project very much at the shoulders. Towards the tail, on the contrary, the body becomes narrow — the conformation being thus exactly opposite to that of the Shanghae. The tail is small, and that of the cock droops. The plumage is very close, firm, and glossy, more so than that of any other breed, giving to the bird a peculiar luster when viewed in the light. The colors vary very much. We consider pure white the most beautiful of all; but the most usual is that well known under the title of brown-breasted red game. The legs are yellow, but quite naked. The head and beak are long, the latter being rather hooked. Comb low and flat, covered with small prominences like warts. Wattles and deaf-ears very small. Eye usually yellow. The whole face and a great part of the throat are red and naked, and the whole expression 'snaky' and cruel. This is not belied by the real character of the breed, which is most ferocious, even more so than Game fowls, though inferior to the latter in real courage."

THE FRIZZLED.

We can find no difference between the "crisp-feathered" and French frizzled fowl. LAYARD says these fowls were first found in Batavia, but TEMMINCK avers they are natives of Southern Asia, and are largely bred and

TRIO OF FRIZZLED FOWLS.

domesticated in Java, Sumatra, and on all the Philippine Islands. They are known by BRISSON as *Gallus crispus* (frizzled fowl,) and as *Gallus pennis revolutis* (fowl with rolled-back feathers) by LINNAEUS. The prevailing color of these birds is white, but there are many specimens variously colored with black and brown. We were highly impressed with their novel appearance, and, as ALDROVANDUS says in his description of them, two peculiarities of the cock attracted our particular attention and admiration. First, that the feathers of the wings had a contrary situation to those of other birds; the side which in others is undermost or inmost, in this was turned outward, so that the whole wing appears inverted; the other, that the feathers of the neck were reflected towards the head, like a crest or ruff, the whole tail feathers turning in the same manner.

As near as we can learn, this variety of fowl does not possess any peculiar advantages over the common barn-yard breed, and is more interesting as a curiosity than valued for any practical purposes. They would undoubtedly thrive in our warm southern far better than in our cold northern climate. The hens make good mothers; they breed freely with all other domestic fowls, and the offspring is prolific without end, the chicks being perfectly hardy, and make a good table fowl, though rather small.

THE DORKINGS.

In years gone by the Dorkings were the favorite fowls in this country, and the only reason we can assign for their degeneracy is the improper care they have received and the continual in-and-in breeding. To rear Dorkings profitably it is essential that a good, long runway should be provided on a clay or gravelly soil for the chicks. They never should be allowed to run on wooden or brick floors. If this is carefully attended to the chickens *will thrive* and grow well, and make hardy fowls.

There are two species of these fowls — the white and the colored Dorkings. The former is the favorite bird of old fanciers, and a writer in the *Poultry* (English) *Chronicle* makes the following remarks on this breed of fowls:—"The *old* Dorking, the *pure* Dorking, the *only* Dorking, is the White Surrey Dorking. It is of good size, compact and plump form, with short neck, short white legs, five toes, a full comb, a large breast, and a plumage of spotless white. They are hardy, lay well, and are excellent mothers."

WHITE DORKING.—We have reared the White Surrey Dorkings for a number of years, and fully coincide with the writer in the *Chronicle*. To our mind, no fowl is more essential to the farm-yard than the *pure* White Surrey Dorking. The first pen of Dorkings we ever experimented with were purchased of Judge S. S. BOWNE, in 1852. His stock was procured from imported fowls of Dr. EBEN WIGHT of Boston, who was at that time the largest breeder of fancy fowls in this country. Our experiments with the Dorking prove them to be fowls not to be despised. They are not early layers, but make up this deficiency in the number and quality of eggs they produce. They are easily fattened, and their flesh is of the very best quality.

In speaking of the weight of the Dorking, the *Practical Poultry Keeper* says:—"It is difficult to give a standard; but we consider that a cock which weighs *less* than ten pounds, or a hen under eight and a half pounds, would stand a poor chance at a first-class show." We have never, in our experience, seen one brought to this weight, not even by high feeding. Our yearling fowls have often been brought to weigh from six to eight pounds.

The practice of crossing Dorking pullets with a game cock is much in vogue, with the object of improving a worn out stock. This, however, would be better accomplished by procuring a fresh bird of the same kind, but not related. This cross shows itself in single combs, loss of a claw, or an occasional red feather, and, what is still more objectionable, in pale yel-

WHITE DORKING COCK.

low legs, and a yellow circle about the beak. These are faults in the Dorking to be avoided by breeders generally.

SILVER GRAY DORKING.—Among the breeds of colored Dorkings which are now attracting attention in this country with fanciers, is the Silver Gray variety. Nearly all authorities aver that this breed is a chance off-shoot from the White Dorking, the breed having been perpetuated by careful breeding. Still, colored birds frequently throw silver-gray chicks, but disappointments are as often sure to follow in breeding for this cross, unless, when obtained, the strain is kept pure for years, as in the case of the Derby Red Game fowls. The only way to accomplish this is to remove all chicks from the pens that do not show the perfect markings of the parent stock.

Mr. HEWITT of Sussex, says the colored Dorkings are decidedly the most useful of all fowls for general table purposes, and a very important point in the consideration of the Gray Dorkings is that they grow rapidly and are in good condition at almost any age, if at all freely supplied with food. The distinguishing colors of the Silver Gray Dorking cock are per-

PAIR OF GRAY DORKINGS.

fectly black breast, tail, and larger tail coverts; the head, neck, hackle, back, saddle and wingbow a clear, pure, silvery white. Across the wings there should be a well-marked black bar, contrasting in a very striking, beautiful manner with the white outer web of the quill-feathers and the silvery white hackle and saddle. The breast of the hen should be of a salmon-red color, passing into gray towards the thighs. The neck a silvery white, striped with black; the back silver gray, with the white of the shafts of the feathers distinctly marked; the wings a silvery or slaty gray, and free from any tendency to redness; the tail a dark gray, the inside nearly black.

Dorkings, like other breeds of fowls, are apt to degenerate very fast from inter-breeding, therefore care should be taken to introduce fresh blood frequently, or disappointments are sure to follow.

Mr. DOUGLAS, an eminent English breeder, says he has found the dark-colored Dorkings the most hardy and heaviest in flesh. He once had a cock weighing fourteen and a half pounds at two years, and several hens at eleven pounds each. He claims that early Dorking pullets will lay all the winter, although not so freely as some other breeds. They lay from thirty-five to fifty eggs before wanting to sit. As mothers, they are perfectly docile, and allow themselves to be handled at will; chickens from other hens may be placed with them, which they will take to at once. These fowls are not classed among the roamers, but are rather of the stay-at-homeativeness birds, therefore are of little trouble to the housewife, and can be easily reared.

FAWN-COLORED DORKING.—A writer in one of the agricultural journals of New England gives the following description of the Fawn-colored and Black breed of Dorkings. He says the fawn-colored bird is made up of a cross between the White Dorking and the fawn-colored Turkish fowl. They are of lofty carriage, handsome and remarkably healthy. The cocks weigh from eight to nine pounds, and the hens from six to seven; they come to maturity quite early for so large a fowl. Their tails are shorter and legs darker than those of other Dorkings; their flesh is fine and their eggs are very rich. It is conceded to be one of the best varieties of fowl known, as the size is readily increased without diminishing the fineness of the flesh.

BLACK DORKING.—The Black Dorkings are said to be of large size, and of a jet black color. The neck feathers of some of the cocks are tinged with a bright gold color, and some of the hens bear a silvery complexion. Their combs are usually double, and very short, though sometimes cupped, rose or single, with quite small wattles, and are usually very red about the head. Their tail feathers shorter and broader than the White variety, and the chicks feather much slower. The legs of the Black are short and black, with the usual five toes on each foot, the bottom of which is frequently yellow. The two back toes are quite distinct, starting from the foot separately; frequently showing an extra toe between the two. This breed commences laying when very young, and lay well during the winter season—the eggs being of a large size. The breed is perfectly hardy, and are good setters and attentive mothers to their young.

THE HAMBURGS.

This breed of fowls is considered a very useful and important denizen of our poultry-yard. We have bred them for years successfully and with little trouble. The hens are inveterate layers, and seldom desire to sit; their propensity for laying being almost continuous from one molting season to another. This is undoubtedly owing to their confined condition in this country; for it is said that when the birds have a free range, they frequently set themselves to the task of incubation with as much diligence as other fowls.

PENCILED HAMBURG.—The penciled Hamburg, which is of two colors, golden and silver, is very minutely and beautifully marked. The cocks do

PAIR OF SILVER-SPANGLED HAMBURGS.

not exhibit the pencilings, but are white or brown in the golden or silver birds respectively. They should have bright double combs, which are firmly fixed upon the head, ending in a point which turns upward; well defined

PAIR OF GOLDEN-SPANGLED HAMBURGS.

deaf ears; taper blue legs, and ample tails. The carriage of the cock is gay and majestic; his shape is symmetrical, and appearance indicative of cheerfulness. The hens, of both varieties, should have the body clearly penciled across with several bars of black, and the hackle in both sexes should be perfectly free from dark marks. These birds are imported in large numbers from Holland to England, from whence we derived the breed; but those now bred in this country are far superior to the imported bird both in size and beauty of plumage.

SPANGLED HAMBURG.—Of the Speckled or Spangled variety, which is

SILVER-SPANGLED POLAND COCK.

becoming a great favorite with many breeders in this country, there are two kinds—the Golden and Silver Speckled. The general color of the former is golden, or orange-yellow, each feather having a glossy dark brown or black tip, particularly remarkable on the hackles of the cock and the wing-coverts, and also on the darker feathers of the breast. The plumage of the hen is yellow or orange-brown, and in like manner being marginal with glossy black. The Silver-Spangled breed is distinguished by the ground color of the feathers being of a silver white, with perhaps a tinge of straw yellow; every feather should, however, be margined with glossy black.

Both of these varieties are extremely beautiful, commanding, as they do, high prices. The hens in all cases proving prolific layers and non-setters.

BLACK HAMBURG.—This is one of the finest varieties of our black fowls—the plumage being of a beautiful black color with metallic luster. They possess the two-fold advantage of being noble-looking birds and exceedingly good layers. On the whole, the Hamburg is a capital fowl, and

SILVER-SPANGLED POLAND HEN.

one which is deservedly highly valued. It has a good, robust constitution, and proves perfectly hardy in almost any climate. Though the eggs produced by this breed are not as large as those of some other breeds, still what they lack in size is made up in the number they produce during the year.

THE POLANDS.

There are several varieties of these fowls in this country, but those possessing the most prominence among breeders are the Silver, Golden Spangled, White, Black, and Black with White top-knot.

SILVER-SPANGLED POLAND.—We have bred this variety for several years, and find it one of the most desirable breeds for the poultry yard, proving with us perfectly hardy and "everlasting layers." The ground color of the plumage of the Silver-Spangled Poland should be a silver white, with well defined horseshoe-shaped black spangles. In the cock, the hackle feathers are white, edged and tipped with black; in the hen, each hackle feather has a spangle on the end; tail feathers clear white, with spangle on the end; the spangles on the wing coverts are large and regular in both sexes, so as to form *two* well-defined bars across each wing. The proper

PAIR OF GOLDEN-SPANGLED POLANDS.

spangle on the breast is all-important. The crest should be full and regular; feathers black at the base and tip, with white between. A few white feathers frequently appear after the second molt, in the very best hens. Ear-lobes small and white; wattles, none, being usually replaced by a black or spangled beard. The weight of the cock is from six to seven and a half pounds, while that of the hen is from four to five and a half pounds. Besides the moon-shaped spangles, many of the birds are shown with laced feathers—*i.e.*, with an edging of black on the outline of the feathers, but thicker at the end. This marking, when perfect, is of exquisite beauty. Dr. BENNET says they certainly rank among the very choicest and most beautiful of fowls, whether considered for their beauty or rarity. The newly hatched chickens are very pretty, creamy white, interspersed with slaty dun on the back, head and neck, marked with longitudinal stripes down the back, with black eyes, light lead-colored legs, and a swelling of the down on the crown of the head, indicative of the future top-knot, which is exactly the color of a powdered wig. At a very early age, they acquire their peculiar distinctive features, and are then the most elegant little miniature fowls it is possible to imagine. The distinction of sex is not very manifest till they are nearly full grown, the first observable indication being in the tail—that of the pullet is carried upright, as it should be, while the cockerel's remains depressed.

PAIR OF WHITE-CRESTED BLACK POLANDS.

GOLDEN-SPANGLED POLAND.—This variety varies in the color of its plumage from a light to a dark golden yellow, laced and spangled with a greenish luster black, and not unfrequently showing some part whitish feathers in their wings, tail and crest. Legs and feet usually bluish, sometimes verging on a greenish color; ear-lobes bluish white.

BLACK POLAND.—The Black Polands are no strangers in this country, they having been bred as long ago as we can remember. In plumage they should be uniformly black (except crest,) although not unfrequently glossed with metallic green, which, in contrast with the deep red wattles and handsome crest of white feathers, gives them a very unique appearance. Their legs are usually dark colored, although through too close breeding, flesh-colored and even yellowish legs will show themselves; but those with darkish legs are to be preferred. Often times the cock will have some whitish feathers in the tail, which by some is thought to be a sure sign of pure breeding.

WHITE-CRESTED BLACK POLAND.—The White-crested Black is of a glossy black color; body short, round and plump; legs shortish and of black or leaden color; full wattles of a bright red; ear-lobes pure white; hackle, saddle and tail have bright reflections; crest is of pure white, regular and full. These birds weigh from five to six pounds.

WHITE-CRESTED WHITE POLAND.—The pure White-crested White Polands are very hardy; have no wattles, but have a well-developed beard in lieu thereof. They, like all Polish breeds, are "everlasting layers," and non-setters. There may be seen occasionally Blue, Gray and Cuckoo Polands, but they are off-shoots, or the result of crossing, and have no qualifications worthy of particular notice.

THE LEGHORNS.

It is said that this breed of fowls was imported from Leghorn, Italy, only a few years since, but has been bred to such perfection in this country that there has been a distinctive breed made, and become, as it were, Americanized. They are scarcely known in England, but are highly prized by American breeders for their many good qualities. They are bred of nearly all colors save black—the White, however, receiving the preference. The imported birds are not inferior to the American standard of excellence. The white variety being similar to the Spanish

PAIR OF WHITE LEGHORNS.

in size and appearance, except in the plumage, which is white, with hackle or neck and saddle feathers slightly tinged with gold. They have proved thus far very hardy birds, suffering from the sudden changes and severe weather of our northern and western climate much less than the Spanish, with which breed many deem them closely allied. They are extremely good layers, and seldom desire to set. The young are easy to rear; they feather up soon, and at the age of six or eight weeks are miniature chickens—that is, perfectly feathered, and as sprightly as many chicks are at four months of age. The hens are considered excellent winter layers, and will lay as large a number of eggs in a year as any fowls known, not excepting the Polands or Hamburgs. They are hardy, medium sized fowls, of a quiet and docile disposition; persistent layers of a rich, meaty egg; pure white color, though in some flocks occasional colored feathers will appear; these should at once be discarded from the pen, if it is desirable to breed the pure white bird. Their legs and skin should be of a yellow

PAIR OF EARL DERBY GAMES.

color. They lay a smaller egg than the Spanish, but mature earlier, and are much superior for the table. The cocks have large single combs, which should stand perfectly erect; full wattles and large, cream-colored or white ear-lobes, extending sometimes upon their face. The carriage of both cock and hen is proud and dignified. The hens have usually large combs, which frequently lop over like the Spanish. From what we have read and seen of this breed of fowls we consider them a great and valuable acquisition to the poultry-yard.

THE GAMES.

The varieties of so-called game fowls are almost innumerable. Many are unworthy of the name or the prefix. A well-bred game cock should be a neat, trim fowl, feathers close and glossy, head small, neck well set on his shoulders, toes lengthy, body erect and straight, strong on thigh, quick in motion, and willing to die for his flock rather than yield to an opponent. Game hens possess the same general qualifications. They should be excellent layers and sitters, and for rearing chicks they are considered superior; they are hardy, strong, and transmit these peculiar traits, as a general thing, to their offspring.

The flesh of the Game fowl is fine and sweet, and is esteemed of a decidedly rich flavor. In this breed almost all shades of feathers are allowable, black-reds perhaps being most common, although jet blacks, pure whites, grays, ginger-reds, spangles, or pied, and various blendings of colors called piles, have their respective admirers, as the fancy of the

PAIR OF BLACK-RED GAMES.

breeder dictates. The breeds also are numerous; those of English, Irish, Mexican, Spanish, Cuban, Malay and other nationalities claiming equal attention with fanciers in their respective localities.

EARL DERBY GAME.—This is an old breed, one which has been given the preference for years, and from which the black-breasted reds undoubtedly originated. The best information that we can gather relative to this breed is that they were originally imported from Knowlsley, Eng., where they have been bred with great care for upwards of one hundred years, in all their purity. The cock is of good round shape, well put together; the head being long, with daw-eyes, long and strong neck; hackle well feathered, touching the shoulders; wings large and well quilled; back short; belly round and black; tail long and sickled, being well tufted at the root—thick, short, and stiff; legs rather long, with white feet and nails, the latter being free from all coarseness. The required "Daw-eye" is that which resembles the gray eye of the jackdaw. Their distinctive features are the white beak, feet, and claws, essential to every bird claiming descent from that illustrious stock. The red Derby Game cock should have a bright red face; breast and thighs coal black; hackle and saddle feathers light orange-red; back, intense brown-red, a depth of color that painters term dragon's blood; lesser wing-coverts maroon colored; greater wing-coverts marked at the extremity with steel-blue, forming a bar across the wings; primary wing-feathers bay; tail iridescent black. It seems a peculiarity in these fowls that one at least of the pinion feathers is marked with white. The sex of the chickens can readily be distinguished when only a few weeks old. The beak, legs and feet are uniformly white. MARTIN remarks that "through the whole catalogue of game fowls the male birds are by far the most conspicuous in plumage;" and this remark proves true in regard to the Derby breed, for wherever mere color has given the name of a class, the markings of the cock explain the reason. The Black-breasted red hens possess little of their consort's brilliancy of feather, though these are of much lighter colors than the red-breasted hen—a fact in strange opposition to the plumage of the respective male birds. BEETON's Poultry Book thus describes the perfect markings of the Lord Derby game hen:—"Head fine and tapering; face, wattles, and comb bright red; extremities of upper mandible and the greater portion of the lower one white, but dusky at its base and around its nostrils; chestnut-brown around the eyes, continued beneath the throat; shaft of neck-hackle light buff; web pale brown, edged with black; breast shaded with roan and fawn color; belly and vent of an ash tint; back and wing coverts partridge-colored; primary wing-feathers and tail black, the latter carried vertically and widely-expanded; legs, feet and nails perfectly white." The carriage of both cock and hen of this breed is upright and dignified. The pugnacious disposition of the cock equals that of any other game bird; and its endurance cannot be surpassed; years agone they were numbered among the best breed of birds for the cock-pit; and for the table they are not surpassed by the sweet and nutritious flesh of the Dorking fowl.

DUCK-WING GAME.—The *pure* Duck-wing Game fowls are the Silver Grays—though there are Yellow or Birchen Duck-wings, but the blood of the first mentioned is much purer than the other variety, and it is considered a much finer, hardier, and more pugnacious bird. The cock should be of a silver gray color; hackle striped, with black underneath, but clear above; back bright silver gray; breast clear, mealy silver gray color; wing crossed with a steel blue bar, the lower part of a creamy white; tail greenish glossy black. The plumage of the hen should be of a silvery blueish gray, thickly frosted with silver; breast pale fawn-color; neck-hackle silvery white, striped with black. The comb and face in both sexes are of a bright red. The legs of the silver gray should be white; eyes red and skin white.

THE YELLOW DUCK-WINGS.—The Yellow Duck-wing Game fowl is of straw or birchen color, with copper-colored saddle; skin yellow, and willow or yellow legs. The cock's breast, in this variety, is always black, while that of the hen is fawn-colored. The weight of the cocks of the Duck-wing variety of game birds varies from four to six pounds, while that of the hens exceeds that of the cocks.

DUCK-WING BANTAMS.—In courage and endurance the Bantams are not behind their larger relatives, and in constitution they are much hardier

PAIR OF DUCK-WING GAMES.

than any other of the Bantam breeds. The plumage of the Duck-wing Bantams is precisely similar to that of the larger breed, from which they were undoubtedly obtained, by long inter-breeding with the smallest specimens. The carriage and form are also similar; but the drooping wing of the Bantam breed is not to be observed in the game variety. In weight the cock does not exceed one and a half pounds, while that of the hen is about twenty ounces. Game fowls can be as easily kept on a "town lot" as any other breed, and with as little trouble. If they are well fed, and proper care taken of them, they are not pre-disposed to roam, but remain quietly at home.

SALMON PILE GAME.—Coloring of hens is a buff or straw color, underlined with white, and has a rich creamy or salmon-colored look; although some specimens are shaded more or less with red or light wine cast. Cocks at maturity are beautiful, and in hackle and sickle featherings would be observed as peculiar to this variety. There are but few fanciers who have shown birds of this variety, to our knowledge, in this country;

they claim for them, however, great excellence, as producers of eggs and for table qualities.

DOMINIQUE GAMES take their names from fowls which are common on the Island of Dominica, and in feathering, especially on cocks, are really very beautiful. They are long and rangy in body, well set up or stationed, high, fine heads, and invariably possess thin single combs, free from tassel or head-feathers, while neck-hackle or shawl is made up of long fringe-like feathers, quite uniformly dotted or penciled—so too of the tail hackles. The hens are quite uniform in feathering, although they have more subdued colors. They are very hardy fowls and most prolific layers. Flesh is yellow, and as in almost all of the game varieties, of fine grain and excellent flavor. This variety of fowl is said to be quite scarce in this country.

GEORGIAN GAME.—This well known variety of fowls came originally from Europe, brought over by a gentleman who was a native of Georgia, and celebrated in his time for the reputation his game fowls made for him in sporting circles South. The breed is now generally recognized by most of the poultry clubs, and ranks high with many leading fanciers. They are claimed to have superior laying and table qualities, hardiness, courage, (and what no one will question who has ever seen them,) beauty of plumage, shape and carriage. They are well calculated to stand the rigors of our northern climate, and must be admirably adapted to our warm and genial southern clime.

THE MALAY GAME.—Mr. DARWIN, in his new work "On the Variations in Animals," claims distinctly that the Malay has been bred for years as a game fowl in India; is noted for its courage and endurance; proves successful in the cock-pits of India and adjacent lands. He says they are a small breed of fowls, and are designated in Europe as the "Indian Games;" but in reality are of the original Malay species of game fowls. Mr. HEWITT says he is "not aware of any variety of fowl so cruel, oppressive, and vindictive as Malays; they are literally the tyrants of the poultry-yard." We bred the red Malay years ago, and found that the cocks evinced such a pugnacious disposition that we were glad to get rid of them. In our experience with this breed we found nothing commendable in them for the amateur or fancier; the hens proving only ordinary layers, while neither the plumage or build of the cock is attractive.

SPANISH GAME.—This variety of game fowl is claimed by some writers to be of English origin. It is more slender in the body, the neck, the bill and the legs, than any other variety, and the colors, particularly of the cock, are very bright and showy. The flesh is white, tender and delicate, and on this account marketable; the eggs are small, and extremely delicate. The plumage is exceedingly beautiful—a clear dark-red, very bright, extending from the back to the extremities, while the breast shows a splendid black color. The upper convex side of the wing is equally red and black, and the whole of the tail-feathers white. The beak and legs are black; the eyes resemble jet beads; very full and brilliant; and the whole contour of the head gives a most ferocious expression.

BROWN-REDS.—This breed of fowls has been long bred perfect in outline, and is considered one of our most desirable game birds. The breast of the cock should be red-brown, shoulders frequently of orange-red; comb and face dark purple; beak also dark; wing-butts dark-red or brown; legs blackish brown, with dark talons; hackle, with dark stripes; thighs like the breast; tail a dark, greenish black, and the wing should be crossed with a glossy green bar. The plumage of the hen should be, as a general thing, of a very dark brown color, and penciled with light brown; neck-hackle dark, golden, copper-red, thickly striped with dark feathers; comb and face much darker than that of the cock. The tail-feathers of the hens should show a slight curve; if they are spurred so much the better.

BLACK-BREASTED RED GAME BANTAMS.

BLACK-BREASTED RED is another breed of game that has its hosts of admirers. The plumage of this bird, as its name signifies, should be of a bright red, deeper on the body than in the hackle. Red eyes denote pure blood—any other colored eye in this breed stamps it as a cross. The cock's hackle is striped underneath, but never above; the comb and wattles bright red; the wings are of the same color in the upper part, and rich red chestnut in the lower, with steel blue bar across; breast bluish-black, with glossy reflections; thighs the same; tail greenish black, without much down at the roots of the feathers; legs are usually willow in color. The hen should be of a rich partridge-brown, with red, fawn-colored breast; reddish golden hackle with dark stripes. There are several other game fowls which have their friends and admirers in this country, such as White,

DUCK-WING GAME BANTAMS.

Black, Gray, Dark Gray and Piles of all colors, but those considere d most merit by breeders are given in their order.

PILE GAME.—The plumage of these fowls should have a proportion of white as one of its compound colors. The cocks of all the various strain of Piles are red and white, yellow and white, in one or other of the shades of those colors. The best Piles are bred by crossing red and white game, but may also be bred from a Pile cock and Pile hens. Some of the best and purest may be bred from a Spangled cock and White hen. The object of the breeder, particularly of show birds, should be to get the colors of the cocks as distinct and as brilliant as possible.

THE BOLTON GRAYS.

This breed of fowls derives its name from having first been successfully raised in and near Bolton, England. They are now found in almost all large poultry-yards in America, as well as in Europe. They are also known in some parts of the United States as the Creole fowl, from the mottled appearance of the hen, whose every feather is delicately marked with alternate bands of black and white, legs and feet a light blue, and very short. The neck-hackle is white. The cock's plumage is different from the above in many respects, his feathers being nearly white. His tail is black, and legs and feet the color of those of the hen, but are much longer. In weight he is less than the hen. Those of our own raising weigh about five pounds to the hen, and about three and a half or four pounds to the cock. One singular peculiarity of the hens of this breed is that they are furnished with *spurs* over an inch in length, while those of the cock are much shorter. The Bolton Grays begin laying early in February, and continue throughout the year. If well fed, they will lay all the year round. Their eggs are below the average size; but what they lack in size is made up in number. As a breed, they are exceedingly hardy, and thrive where many breeds would perish. They are not good sitters, and their eggs must be set under some other fowl. They are never inclined to wander away from their coops.

THE BLACK SPANISH.

This is one of our best black breeds of fowls, laying as they do a large sized and meaty egg. The cock should carry himself very stately and upright, the breast well projecting, and the tail standing well up. The sickle-feathers should be perfect and fully developed, and the whole plumage a dense jet black, with glossy reflections in the light. The hen should be equally dense in color, but is much less glossy. Any white or speckled feathers, which now and then occur, are fatal faults. The legs should be blue, or dark lead-color; any approach to white is decidedly bad. The legs of both sexes are long, but the fowl should nevertheless be plump and heavy. The comb must be large in both sexes, and of a bright vermillion color. That of the hen should fall completely over on one side; but the cock's comb must be perfectly upright. The indentation also must be regular and even, and the whole comb, though very large, quite free from any appearance of coarseness. Any sign of a twist in front is a great fault. The most important point, however, is the white face. This should extend as high as possible over the eye, and be as wide and deep as possible. At the top it

WHITE AND BLACK SPANISH FOWLS.

should be neatly arched in shape, approaching the bottom of the comb as nearly as possible, and reaching sideways to the ear-lobes and wattles, meeting also under the throat. In texture the face ought to be as fine and smooth as possible. The ears are large and pendulous, and should be as white as the face. Any fowls with red specks in the face are considered very faulty.

WRIGHT says the other principal varieties of Spanish fowls are Minorca or Red-faced Black, the White, the Blue or Andalusian, and the Ancona, Gray, or mottled breed. The plumage of the White Spanish is of snowy whiteness and resembles somewhat the White Leghorn. We found the Spanish in Western New-York to be very susceptible to disease, and great care was necessary to keep their combs and wattles from being frost-bitten. In a warm climate, we dare say, the Spanish as a class cannot be beaten. They do not do well in confinement; they are predisposed to roam; such has been our experience with them.

THE PLYMOUTH ROCKS.

This breed of fowls we hardly think is known outside of the New England States. It is said the Plymouth Rock is produced by crossing a Cochin China cock with a hen, a cross between the Fawn-colored Dorking, the great Malay, and the Wild Indian. The cock has been bred to stand, at a year old, from twenty to twenty-five inches high, and weigh from eight to ten pounds; the pullets from six to seven pounds each. Generally speaking, the pullets are very early layers; commencing at five months of age and continue to lay until the molting season. They lay a medium sized egg, of a rich and reddish-yellow color. The plumage of these fowls is very rich and variegated, showing off in the sun the most brilliant hues. The cocks are usually of a beautiful red or speckled color, and the hens of a darkish brown. Some of the colors thrown by this breed are not dissimilar to the Dominique fowl. They have very fine flesh, and are fit for the table at an early age. The legs are quite large, and usually blue or green, but occasionally yellow or even white, and frequently having five toes upon each foot. Some of the varieties have the legs occasionally slightly feathered. They have large single rose-colored or red combs and wattles; cheeks are rather large; tails stout and short, and very small wings in proportion to their bodies. The chicks are quite hardy and have the same uniformity in size and appearance as those of the pure bloods of primary races. The hens make good mothers and close setters.

THE JERSEY BLUES.

These fowls were bred to some extent twenty years ago, and were deemed by many a very valuable breed. In 1855 we bred them for a time, but finding them possessed of no superior qualities, discarded them for the White Shanghaes. The color of the Jersey Blue is a light blue, sometimes approaching a dun; the tail and wings rather shorter than those of the common fowl; the legs are generally black, though we have bred them of a dark blue color, somewhat lightly feathered. They proved with us perfectly hardy, but were not prolific egg-producers. The cocks at a year old weigh from six to eight pounds, while the hens weigh from five to seven pounds; flesh rather coarse, stringy and unnutritious.

THE FRENCH BREEDS.

The French breeds of fowls lately introduced in this country may be classed as first, the Houdan, second, the Creve-Cœurs, and third, the La Fleche. The de Bresse, du Mans, de Breda, Courte Pattes, and the more ornamental, as the Padoue, Chamois, Hermines and Hallandais are known only in this country by name. The Houdan and Creve-Cœurs are bred to some considerable extent in this country, but the La Fleche has not as yet been raised with sufficient productiveness or hardihood among us to be

PAIR OF HOUDANS.

fully appreciated. We think they may be successfully bred in our warm and genial southern climate, for when once reared they stand second to none as a table fowl, or layers of very large eggs.

The Houdans.—These birds derive their name from a village in France, where they were originated. They are held in as high estimation in France as the Dorkings are in England. This breed needs no inducement to increase and multiply, for they are easily reared and fattened, and being constant layers of good sized eggs, with the quality of the flesh fine, they are a desirable fowl for the poultry keeper to breed. They possess vivacity tending to wildness, bearing confinement and enjoying liberty with spirits that never flag; they are "bright as a flower and upright as a bolt." At shows it is required of them to possess the fifth toe, and perpetuate the useless monstrosity of their semi-original, the Dorking, from whom and the silver Padoue they are doubtless descended. Color rocky white and black; an even speckled proportion of each preferred. Occasional stained feathers appear in the purest blood, but red ones tend to disqualify. The

PAIR OF CREVE-COEURS.

head is crowned with a fierce tuft, and on the front rises a horned or double-leafed comb, the center having the appearance of an ill-shaped long strawberry. The whiskers and beard are striking, growing well up on the face of both cock and hen. The legs are spotted leaden grey. The hen's crest should be thick and full, showing as little comb as possible. The *coup d'œil* of a company of these birds is most brilliant, and it is to be hoped that their weight (as yet but moderate) may in time approximate to that of the Dorking, whose contour and volume they imitate.

THE CREVE-CŒURS.—The Creve-Cœurs are of bold mien and grave aspect, with black plumage glistening with green; crested heads lighted up with crimson-colored, antler-like comb. Their contour strikes the observer with the idea of usefulness and dignity, nor is the notion illusory; short-legged, heavy, with little offal, much aptitude to fatten, and (save when

A PAIR OF LA FLECHE FOWLS

very newly imported) sufficiently robust, steady egg-producers, and growing to adolescence with moderate care, they merit our careful regard. To go into detail, the crest of the cock should be formed of lancet-like feathers, fairly raised; not too regularly placed; the comb should be full and large, regularly irregular, with pendent and long wattles, voluminous and deep beard, thick plumage, especially on the breast, full tail, horizontal back, short legs of a leaden-blue color, firm claws. The crest of the hen is more round, soft, and thick; the less appearance of comb or wattles the better. Though a perfect blackness of color is required in both sexes, the very best specimens will show a white feather or two in the crest as age advances, but red or straw colored streaks are not tolerated among the aristocracy of the breed. They came originally from Normandy, principally the county d'Ange, where lies the pretty village of Creve-Cœur.

THE LA FLECHE.—The La Fleche is a Malay in hight, a Spanish in color, and a Dorking in size. It possesses a firmly knit, angular body,

PAIR OF GUELDERS.

poised proudly on long, nervous, strong limbs, not showing the bird's complete size, owing to the closeness of his feathering; a little spike of feathers is placed behind the comb, which appears as a double horn; the aspect of a rhinoceros is given to the head by a dwarf protuberance between the nostrils, which are much expanded; very long pendant wattles; large opaque white ear-lobes, expanding in a cravat; gently curved strong beak; neck-hackles long and fine, reflecting, as well as the feathers of breast, wing, and upper tail, violet and green black; color not so bright below; claws especially strong; legs slaty blue, and in age leaden gray. Hen identical, but somewhat smaller, with less comb, ear-lobe and wattles. She grows for twelve months, the cock for eighteen; and it is this continuity of growth that enables the breeder of this superb table fowl in France to obtain a golden price for his spring lots. The young feather slowly. They are raised on the commons of the arrondissement of La Fleche.

THE GUELDERS.

This variety of fowls is as yet very little known in this country, but what we can learn of their qualities, from those who have experimented with them, we are led to believe, after they have become acclimated, they will prove a very desirable breed. These birds were first found in Holland and Belgium, and are known in those countries as Guelderlands, being so called after a province in Holland, lying south of the Zuyder-Zee. There are White, Black and Cuckoo Guelders bred in this country. A gentleman of our acquaintance, who has bred these birds for the last two or three years, considers them superior to any of the French fowls, and in some respects prefers them to Houdans.

The Guelders are of medium size, with full, prominent breasts, and large flowing tails. Their peculiar characteristics are in the head, which is destitute of either feathers, crest or comb; the latter is very peculiar in shape, being hollowed or depressed instead of projecting, with two prominent spikes on each side of the back of the comb. To breed them to the standard, they should not have any comb whatever, except the two little spikes projecting. Cheeks and ear-lobes red; wattles red, and in the cock very long and pendulous. The beak in the White should be of a milk-white color. The thighs well furnished and vulture hocked, and the shanks of the legs feathered to the toes, though not heavily. The plumage is close and compact, resembling very much that of the Game fowl, which makes them appear, in size, much smaller than they really are; the color of the plumage in one is pure white, and in the other pure black. To produce the Cuckoo-colored bird a Black Guelder cock should be placed with a White Guelder hen. By this cross Cuckoo-colored birds of a beautiful variety have been thrown. It is said that the Guelders, thus far, have withstood our cold and changeable northern climate equally as well as the Asiatic breeds; have proved very hardy and less susceptible to sickness than any other class of fowls. They are small eaters, lay a large, smooth-shelled egg, and seldom desire to sit. As egg-producers, especially in cold weather, it is asserted they are not surpassed even by the Leghorn, and lay throughout the year more eggs than any other breed of fowls. Their flesh is nearly as delicate and juicy as that of the Houdan. The chicks are easily reared, under ordinary circumstances, and feather up very quickly.

THE DOMINIQUE.

In speaking of these fowls, Mr. BEMENT says "they are distinguished as Dominique by their markings and their color, which is generally considered an indication of hardiness and fecundity. They are by some called 'Hawk-colored fowls' from their resemblance to the birds of that name. We seldom see bad hens of this variety, and, take them all in all, we do not hesitate in pronouncing them one of the best and most profitable breed of fowls, being hardy, good layers, careful nurses, and affording excellent eggs and first quality flesh." Dr. BENNETT, in his description of the Dominique, says:—"The prevailing and true color of the Dominique fowl is a light ground, undulated and softly shaded with a slaty blue all over

DOMINIQUE COCK.

the body, (as indicated in the portrait of the cock herewith given,) forming bands of various widths; the comb of the cock is variable, some being single, while others are double—most, however, are single; the iris, bright orange; feet and legs are bright yellow or buff color; bill the same color as the legs." *Browne's Poultry Yard* remarks that they are not only good layers, sitters and nurses, but that "their beautiful appearance, when in full plumage, is quite an acquisition to the farm-yard or lawn." Taken all in all we consider them one of our very best breeds of native fowls, and one that alters little by in-and-in breeding.

THE BANTAMS.

Since the first introduction of the Bantam breed of fowls they have ramified into many varieties, none of which are destitute of elegance, while

TRIO OF SILVER SEBRIGHT BANTAMS.

some, indeed, are remarkably beautiful. All are, or ought to be, of small size, but lively and vigorous, exhibiting in their movements both grace and

TRIO OF PEKIN OR COCHIN BANTAMS.

stateliness. The feather-legged Bantam is very remarkable for the *tarsi*, or beams of the legs, being plumed to the toes with stiff, long feathers which brush the ground. The black-breasted reds are considered fine birds. They are red in color, with a black breast and single dentated comb. The *tarsi* are smooth, and of a dusky blue. When this breed are bred pure, it yields in spirit and courage to none, and is, in fact, a game fowl in miniature, being as beautiful and graceful as it is brave. A pure white Bantam is also a beautiful bird, and as courageous as it is beautiful. The Golden and Silver Sebrights, the Nankeen and Pekin Cochins are also remarkably handsome birds, as are also the Black Bantams.

GOLDEN AND SILVER SEBRIGHT BANTAMS.—The plumage of the Golden Sebright is of golden color, and the Silver Sebright of a silver white, with a glossy jet black margin; the cocks have the tail folded like that of the hen, with the sickle feathers shortened or nearly straight, and broader than usual.

GOLDEN SEBRIGHT BANTAMS.

BLACK BANTAMS.—The plumage of the Black Bantam is a uniform black in color, resembling that of the Black Spanish; tail of the cock arched; legs short, dark blue or black, and perfectly clean; comb a bright red; ear-lobes white; face red. Hen not to exceed eighteen and the cock twenty ounces.

WHITE BANTAMS.—The plumage of the White Bantam is pure white, with legs white and well feathered. They should not exceed two pounds the pair.

PEKIN OR COCHIN BANTAMS.—This most remarkable of all the numerous breeds of Bantams was first introduced in England in 1862 or '63, and one or two pairs have been shown in this country. It is said the original progenitors were stolen from the Summer Palace, at Pekin. They partake somewhat of the habits of the Cochin Chinas, and resemble Buff Cochins very much in color and form, possessing the feather-leg, abundant fluff, presenting, as the engraving shows, a most singular appearance. To breed them perfect birds in this country, will require great skill; still, by being crossed with other breeds of feather-legged Bantams, to introduce fresh blood, and then breeding back to the pure strain, may have the desired effect. The Pekin Bantams are very tame, the hens are good sitters and mothers; the males even take a share in brooding the chicks. Their novelty will undoubtedly make them great pets among bird fanciers.

THE JAPANESE BANTAM is said to have been imported from Japan. They are very short legged, and have a large single comb. In color some are mottled; others have a pure white body, with a glossy, jet-black tail. This variety is very pretty. As a whole, the Bantams, though small, are not without their good qualities.

THE SILKY.

This variety of fowls, as we learn from the *Practical Poultry Keeper*, possesses two distinct peculiarities. The webs of the feathers have no adhesion, and the plumage is therefore "silky," or consisting of a number of single filaments, which makes the bird appear much larger than it really is,

PAIR OF SILKY FOWLS.

the actual weight of the cock being generally under three pounds, and of the hen about two pounds. The color is usually pure white, but other colors are occasionally seen. The second peculiarity is the dark tint of the bones and skin, from which the name of "negro" fowls is derived. The skin is of a very dark violet color, approaching to black, even the comb and wattles being a dull dark purple. The bones also are covered with a nearly black membrane, which makes the fowl anything but pleasant to look at upon the table; but if the natural repugnance to this can be overcome, the meat itself is white, and very good eating; indeed superior to that of most other breeds. The plumage is often so excessively developed as to give the birds a most grotesque appearance. Our illustration is not in the least exaggerated, and is a good representation of many specimens of the breed. The comb varies in shape; but a Malay comb is best. There is generally a small crest on the top of the head. The legs are mostly well feathered to the ground, and often have five toes; but neither point is universal. The sole value of the Silky fowl is as a mother to Bantams, or other small and delicate chickens, such as pheasants or partridges. For such purposes they are unequaled, the loose long plumage affording the most perfect shelter possible. They are, of course, peculiarly susceptible to cold or wet, and have no other value than that stated, except from their singular and not unornamental appearance.

THE BLACK JAVAS.

This species of birds are said to be among the most valuable breeds of this country, and are frequently described as Spanish fowls. Their plumage is of a black or dark auburn color; legs large and thick; single comb and wattles. They are prolific layers, their eggs being large and as well flavored as those of the Black Spanish. They are a perfectly hardy breed and easily reared.

OLD FARMER FOGY'S FOWLS.

TURKEYS—MANAGEMENT AND DIFFERENT BREEDS

PECULIARITIES OF THE TURKEY.

THAT the turkey has some singular peculiarities in its nature cannot well be gainsayed. Among them may be mentioned its uncommon tenderness when young, and its unqualified hardiness when full-grown. Nothing in the poultry yard is so tender, delicate, and so easily destroyed when first hatched as the turkey. It is easily chilled, past recovery, by cold or storms, and yet, when full-grown, it will endure some of the most severe and pelting storms of mid-winter. We have seen them roost high on the apple trees, during a fierce north-easter, with the snow and ice collecting upon their heads, apparently unconcerned about shelter or protection.

THE REARING OF TURKEYS.

The rearing of turkeys should be one of *the* duties of our farm-house wives, for the turkey is a part of our rural and domestic economy. In our opinion, no farm yard seems complete without having therein a few turkeys commingling with other fowls, for they, next to the common fowl, are the most useful and valuable of our domestic birds; still, to rear them successfully requires patience as well as great care in the management of their young.

HATCHING OF TURKEYS.

To rear a brood of turkeys with any kind of certainty of success, the eggs should be placed under the common barn-yard fowl, or, perhaps, as the Brahma makes a good mother, a hen of that breed will answer as well. We would advise, by all means, not to have the eggs set under the hen turkey—though they are inveterate sitters, they are poor mothers, and it is ten chances to one if they will prove successful, with their roaming habits, in rearing their young. The mother never proves a good provider for them; she never scratches for her young like the hen, but generally leaves them to shift for themselves as soon as hatched. The young, at the moment of their birth, give no sign of seeking their food,—but, being reared by a common hen, whose instincts lead her to scratch and peck for her chicks— the young turkeys soon learn to imitate her example, which determines them to pick up their food and keeps them from starving to death, as they naturally do when left to be provided for by the hen turkey.

TREATMENT OF THE YOUNG.

Turkeys, when young, are quite tender, and need generally more than the "slip-shod" or "make-shift" attention awarded them by many farmers. The first and most essential thing after hatching is to keep them in a dry and warm location. It usually takes from thirty to thirty-two days for the eggs to hatch. As they are hatched the hen or hen turkey, in which ever case it may be, should be placed in a coop with her young brood. We should recommend the "rat-proof" coop to all breeders. For the first three or four weeks after hatching, great care should be taken by the breeder to keep them from the scorching sun, drenching rains, and the heavy morning and evening dews; and this is why the young should be placed in the "rat-proof" coop—that they may be kept dry. Moisture, internal or external, is generally certain death to chickens; cleanliness of the coops should be rigorously observed; dry, gravelly land is the most proper place to keep them on; avoid all grass-plats with the movable coop. The chicks should never be allowed to leave the coop in the morning until the dew is off the grass; be sure to coop them in wet and unpleasant weather. The *American Poulterer's Companion* suggests that as soon as the young ones are removed from the nest, they be immersed in a strong decoction of tobacco, taking care, of course, that the fluid does not enter the mouth or eyes of the chick, and repeat the operation whenever they appear to droop.

THE CRITICAL PERIOD OF THEIR LIVES.

At two periods of their lives, young turkeys need more care than at others. The first is about the third day after they are hatched; and also when they throw out what is termed the "red head," which they do at six weeks of age. This is a very critical period for young turkeys, much more so than at the period of molting; at this time, therefore, their food must be increased, and rendered more nutritious, by adding boiled eggs, wheaten flour, or bruised hemp seed. The English breeder succeeds well by feeding his brood a "mush," made of equal parts of cooked oat and barley meal. This crisis once passed, the birds may be regarded as past danger, and exchange the name chicks for that of turkey poults, and are considered as fairly "toughened."

PREPARATION OF FOOD FOR THE YOUNG.

As we have said before, great care should be exercised in the preparation of their food. Do not feed slop food of any kind. Many breeders feed loppered milk, but that should be scrupulously avoided; it should not be fed under any consideration. Sour milk, *boiled* to a thick curd, is good, mixed with *cooked* Indian meal, seasoning the same occasionally with *black* pepper. They should be fed often, and made to eat up clean what food is given them before repeating the feeding. The food should be thrown on the ground—

not in a trough—so that in picking up their food the gravel that adheres to it will aid their digestive organs to perform their functions. Never feed Indian meal in an *uncooked* state, for it is liable to bake in the crop, causing death in a very short time. Water should be placed in *shallow* dishes, or old tin pie-pans, near the coop, so that the young can satisfy their thirst whenever inclined. At six weeks or two months old the young turkeys may, as a general thing, be considered out of danger from over-feeding, etc., and should then be fed cracked corn, boiled potatoes, refuse from the table, buckwheat, and fresh boiled meat, occasionally, in small quantities.

TO FATTEN TURKEYS.

In regard to fattening turkeys on charcoal, a writer in the *Germantown Telegraph* says:—"I have recently made an experiment, and must say that

DOMESTIC TURKEY.

the result surprised me, as I had always been rather skeptical upon the subject. Four turkeys were confined in a pen, and fed on meal, boiled potatoes and oats. Four others, of the same brood, were also at the same time confined in another pen, and fed daily on the same articles, but with one pint of very finely pulverized charcoal mixed with their food—mixed meal and

PORTRAIT OF A BLACK BRONZED TURKEY.

boiled potatoes. They had also a plentiful supply of broken charcoal in their pen. The eight were killed on the same day, and there was a differ-

WILD TURKEY.

ence of one and a half pounds each in favor of the fowls which had been supplied with the charcoal, they being much the fattest, and the meat greatly superior in point of tenderness and flavor."

THE WILD TURKEY.

THE plumage of the wild turkey is generally described as being compact, glossy, with metallic reflections; feathers double, as in other gallinaceous birds, generally oblong or truncated; tips of the feathers almost conceal the bronze color. The large quill coverts are of the same color as the back, but more bronzed, with purple reflections. The lower part of the back and tail coverts is deep chestnut, banded green and black; the tail feathers are of the same color, undulatingly barred and minutely sprinkled with black, and having a broad, blackish bar toward the tip, which is pale brown and minutely mottled; the under parts duller; breast of the same color as the back, the terminating black band not so broad; sides dark-colored; abdomen and thighs brownish-gray; under tail coverts blackish, glossed with brown, and at the tips bright reddish-brown. The plumage of the male is very brilliant; that of the female is not so beautiful. When strutting about, with tail spread, displaying himself, this bird has a very stately and handsome appearance, and seems quite sensible of the admiration he excites.

THE DOMESTIC TURKEY.

The varieties of the domesticated turkey are not very distinct. There seems to be a question in the minds of ornithologists whether the domestic turkey, so called, is actually a second and distinct species, or merely a variety of the wild bird, owing its diversity of aspect to circumstances dependent on locality, and consequent change of habit, combined with difference of climate and other important causes, which are known in the case of animals to produce such remarkable effects.

THE WHITE TURKEY.

The white turkey is a most beautiful bird, and is supposed by some to be the most robust and easily fattened of our domestic turkeys; but this, from what we have been able to learn upon the subject, is a grave error, they proving, on the contrary, very delicate and hard to rear. But when fattened and killed they dress most temptingly white for the market, and their flesh, when brought to the table, is rather more delicate than that of the common variety.

THE BRONZED BLACK.

This is undoubtedly the finest and strongest bird, resembling as it does, as closely as possible, the original stock, and looks not dissimilar to the wild bird, and next to that weighs the heaviest, fattens the most rapidly, and can be reared with much less trouble than any other variety. We have seen a turkey of this species shown at the New York State Poultry Exhibition that was enormous in size; he weighed upwards of thirty-six pounds. Some turkeys we have seen are of a coppery tint, some of a delicate fawn-color, while others were parti-colored, and gray and white. These are, however, regarded as inferior to the Bronzed-Black, or Black, as their color indicates something like degeneracy of constitution, if not of actual disease.

THE CRESTED TURKEY.

A specimen of this turkey, the only one, we believe, ever exhibited in this country, was shown at the New York State Poultry Show in 1869, and attracted considerable attention; so much so, that we have deemed it not out of place to give an engraving of the head, showing the crest, in these pages, with what description we are able to gather of the same from eminent writers on natural history:—"Amongst the old writers on the natural history of birds," says TEGETMEIER, "are to be found references to a

CRESTED TURKEY.

singular breed of turkeys that were furnished with full crests of feathers." Thus ALBIN, in his "Natural History of Birds," published in 1738, describes a single specimen, belonging to a Mr. CORNELLYSON of Chelmsford. He wrote as follows:—"The back and upper sides of the wings are of a dusky, yellowish brown, the breast, belly, thighs, and under sides of the wings white, the feathers on the lower part of the belly and thighs were edged with black; the tail white, the extreme feathers of which were scalloped near the ends with black, the next circular row scalloped with a dusky yellow; the legs flesh color, having only the rudiments of spurs; the claws dusky."

TEMMINCK, in his "*Pigeons of Gallinaces*," published at Amsterdam in 1813, says:—"The crested turkey is only a variety or sport of nature in this species, differing only in the possession of a feathered crest, which is sometimes white, sometimes black. These crested turkeys are very rare. Mademoiselle BACKER, in her magnificent menagerie near the Hague, had a breed of crested turkeys of a beautiful Isabelle yellow, inclining to chestnut; all had full crests of pure white."

The Rev. E. S. DIXON, in his work entitled "The Dove-cote and the Aviary," quotes the above passage from TEMMINCK, and another from the work of Lieut. BYAM, descriptive of a race of crested wild turkeys in Mexico. The extract from Mr. BYAM I will not quote, as it is quite evident that the bird described by him was not a turkey, but a curassow. The conclusion that Mr. DIXON arrived at was, that there must have been a wild race of crested turkeys from which the crested birds described by ALBIN and TEMMINCK had descended. I need hardly state that there is not the slightest possible foundation for such an opinion, nor for believing in the existence of wild crested fowls, which is also maintained by the writer. Crested turkeys are a variety, not a species; but it is singular that a variety that was so much admired many years since should have passed out of sight, at least so far as Europe is concerned."

It is singular that this particular variety of an American species should now be utterly unknown in its native country, lost entirely in Europe, and only recovered from Africa. When could the breed have been taken there, and how came it to be preserved among the semi-savage tribes of the interior, while it was lost to the civilized races of Europe? Of the origin of this crested breed nothing is now known, but those who are acquainted with the theory of analogous variation, as propounded by DARWIN, will have no difficulty in understanding how such a breed could originate, seeing that several allied genera of crested birds, such as Pavo, Lopophorus, etc., exist.

DUCKS—THEIR VARIETIES AND MANAGEMENT

CAN THEY BE KEPT WITH PROFIT?

ANY calculation as to the return to be expected by those who keep ducks, says an experienced breeder, depends entirely upon the possession of a suitable locality. They are most likely to be kept with profit when access is allowed them to an adjoining marsh, where they are able in a great measure to provide for themselves; for if wholly dependent on the breeder for their living they have such ravenous, insatiable appetites that they would soon, to use an emphatic phrase, "eat their heads off." No description of poultry, in fact, will devour so much or feed so greedy. But certain moderate limits are necessary for their excursions, for otherwise they will gradually learn to absent themselves altogether, and acquire semi-wild habits, so that when they are required to be put up for feeding or immediate sale, they are found wanting. Ducks, too early allowed their liberty on large pieces of water, are exposed to so many enemies, both by land and water, that few reach maturity; and even if some are thus fortunate, they are ever after indisposed to return to the discipline and regular habits of the farm-yard. They may be kept in health in small enclosures, by a good system of management, though we fear not with profit, which is the point to which all our advice must tend. There is no doubt that

DUCKS MAY BE MADE PROFITABLE AS EGG-PRODUCERS,

but the quality of their eggs and the extra labor required to obtain them—for unless they are got up every night and confined, they will drop their eggs carelessly here and there, where they will not be found—will not allow them to compete with the hen in that capacity. Besides, a duck lays when eggs are most abundant, while hens' eggs may be procured at all seasons. The following remarks on rearing and feeding the young are from the pen of the late C. N. BEMENT:—"The

BEST MODE OF REARING DUCKLINGS

depends very much on the situation in which they are hatched. On hatching there is no necessity of taking away any of the brood, unless some accident should happen; and having hatched, let the duck retain her young upon the nest her own time. On her moving with her brood, prepare a coop and pen upon the short grass, if the weather be fine, or under shelter, if otherwise; keep a wide and shallow dish of water, often to be removed, near by them.

THEIR FIRST FOOD

should be crumbs of bread, moistened with milk; curds or eggs boiled hard and chopped fine, are also much relished by, and are good for them. After a few days, Indian meal, boiled and mixed with milk, and if boiled potatoes and a few chives or lettuce chopped fine be added, all the better. All kinds

AYLESBURY DRAKE.

of sopped food, buckwheat flour, barley meal and water, mixed thin, worms, etc., suit them. As soon as they have gained a little strength, a good deal of pot-herbs may be given them, raw, chopped fine, and mixed with a little bran soaked in water, barley and boiled potatoes beat up together.

REASONS WHY THEY ARE USEFUL.

They are extremely fond of angle-worms, grubs, and bugs of all kinds, for which reasons it may be useful to have them run in the garden daily. All these equally agree with young ducks, which devour the different substances they meet with, and show, from their most tender age, a voracity which they always retain. No people are more successful in rearing ducks than cottagers, who keep them for the first period of their existence in pens two or three yards square, feeding them night and morning with egg and flour, till they are judged old enough to be turned out with their mother to forage the field. It is necessary, to prevent accidents, to take care that the ducklings come regularly home every evening; and precautions must be taken, before they are permitted to mingle with the old ducks, lest the latter should ill-treat and kill them, though ducks are by no means so pugnacious and jealous of new-comers as common fowls uniformly are."

OUR PERSONAL EXPERIENCE.

In 1862 we tried the experiment of rearing ducks without having the water facilities said to be necessary to make our undertaking successful. Against the advice of breeders we bought, of the common variety, one drake and three ducks in the fore part of February, placed them in our back yard and let them run with the rest of our fowls; fed them regularly, (as we do all other fowls,) three times a day, and having placed at their command or convenience at all times an eight-quart basin full of water. We did not coop them with our other fowls; understanding they would do

ROUEN DRAKE.

better in dark coops or roosts, we therefore made for them two tight tent coops of rough boards, with small, open doorways in front in the most secluded place we could find in the yard, between a couple of trees and surrounded with shrubbery. The three ducks commenced laying about the last of February, and continued laying pretty regularly until the latter part of August or first of September. In April we set a hen on thirteen ducks' eggs, which brought off twelve young ducks. We did not set any ducks, but continued to use hens for that purpose, and at the close of the season were rewarded with a flock of sixty-eight young ducklings, which brought in the fall, when well fattened, from eight to nine shillings per pair, saying nothing of the large number of eggs used for culinary purposes in a family of ten persons.

REARING DUCKS WITH HENS.

In rearing young ducks with hens we placed near the coops, which were always located in the vicinity of the pump, a small pan or water-tight box sunk in the ground to receive the waste water from the pump, which answered the purpose as well as if they were given a pond of water to swim in, and fretted the hen-mother much less. In fattening them, we gave them plenty of boiled potatoes, mixed with cooked Indian meal, made into a pudding. We fed but little corn or oats. They paid us well for our undertaking, as they doubtless would, if the experiment were tried on a larger scale.

THE AYLESBURY.

This variety of aquatic fowls derive its name from the town of Aylesbury, England, and is highly prized by breeders in this and other countries, on account of the many good qualities which it possesses. They are large, possess excellent table qualities, and are very prolific layers. As for beauty, we do not think a flock of pure white Aylesburys can be equaled. The first importation of this breed was made about the year 1854, we believe, by JOHN GILES, of Woodstock, Conn. The pure bred bird has plumage of unspotted whiteness; a pale, flesh-colored bill; a dark, prominent eye, and orange-colored legs. Dr. BENNETT says:—"The weight of the adult Aylesbury duck should at least average, if properly fed, from ten to twelve pounds the pair. Instances, however, have occurred where the drakes have come up to eight pounds and upwards, and would in all probability, if fattened, reach ten pounds each. They are very prolific layers. From two of these ducks three hundred eggs have been obtained in the course of twelve months, in addition to which, one of them sat twice, the other only once, the three nests giving thirty young ones. The eggs vary in color, some being white, while others are a pale blue. As a further recommendation for them, in an economical point of view, it is argued that their consumption of food is less than that of the common duck; and another advantage may be found in their comparative silence from the continuous 'quack, quack, quack,' of the latter bird. They also attain greater weight in less time; and, from their superior appearance when plucked, are a far more marketable article."

The carriage of the Aylesbury duck is more upright than that of the Rouen, and from its great powers of locomotion the bird is by no means addicted to such stay-at-home habits as the latter. It is not uncommon to see the bill of these ducks turn black or become stained with dark spots as they advance in life. This disfigurement has been greatly commented upon, but no definite conclusion arrived at, still many are disposed to regard it as hereditary. It is averred that the Aylesbury, being a lighter breed, are better sitters and nurses than the Rouens, after the experience of two or three years. The purchaser should be careful not to confound the Aylesbury with another breed of white ducks bred in this country, that

CRESTED DUCK.

were originally imported from Holland; the chief merit of which consists in their incessant quacking; and is termed the "Call Duck." The White Call Duck has a yellow, orange-colored bill, while it should be recollected the bill of the Aylesbury should be flesh-colored.

THE ROUEN.

The Rouens are extensively bred in France and England, as well as in this country, but it is asserted by some writers, were originally from France,

TRIO OF MUSK OR BRAZILIAN DUCKS.

and take their name from the city of Rouen, on the river Seine. TEGET-MEIER, who is at present acknowledged authority on "feathery subjects," seems to think differently as regards the origin of this variety. He says:—"I have been a breeder of this duck many years, and take much interest in their history, but could never discover that Rouen was especially famous for this breed of birds. On the contrary, from inquiries made of poultry fanciers and others who have visited that locality, I have found that these birds were not reared there as a domestic fowl, nor did they abound in a wild state in that district. As to the application of the term, it is most probable that it is a corruption of the word roan, or 'gray duck,' and the animal is not without some claim to the cognomen. As to the origin of the breed, there can be little doubt that it has been bred from the Mallard, enlarged and improved by care and good feeding, and corresponds precisely with it in every respect in the details and markings of the plumage. The markings found also in the wild species are considered as the *criteria* of perfection by poultry fanciers and judges, at the present day, which proves much more than any facts I might advance." This breed of ducks is highly esteemed by breeders for their large size and deliciousness of flesh. They are prolific layers, their eggs being very large, and much valued in England, it is said, for culinary purposes. It may be imagined, from their large size, that they must consume much more food than our smaller variety of ducks; on the contrary, they are remarkable easy keepers, and require less food than the common duck. They commence laying when quite young, are perfectly hardy and not disposed to roam from the immediate vicinity of their yards. There is but little difference in size between the duck and the drake, and in many instances the former will outweigh the latter. They are not good mothers, and their eggs should be placed under a hen to insure success in raising the young. The color of the Rouen *drake* is as follows:—Bill inclined to green, the nail and around the nostrils being black; head and neck, as far as the white collar, which should be very distinct, iridescent green; throat and breast claret-brown; back scapulars, and thighs gray, with minute wavy dark lines at right angles to the shaft of the feather; tail brown, with the outer edge of the feathers white, forming a broad margin of that color, the three center feathers being curled; primaries brown; secondaries the same, with a bar of bright steel-blue forming the speculum, the band of black, the extremities being tipped with white; lesser wing-coverts rich brown; greater wing-coverts the same, with a narrow white margin; under part of the body gray, with the same wavy dotted lines as on the back; legs and feet orange. The plumage of the *duck* is of a rich brown color, every feather being marked more or less with black; bill, legs and feet dusky; irides in both eyes are of a light-brown color. The body of the *ducklings*, when first hatched, is of a yellowish-brown color, and remains so until they are in perfect feather.

THE CAYUGA BLACK.

This is a variety of our duck tribe well worthy of cultivation, and the best of the dark ducks. It originated on Cayuga Lake, one of our most beautiful little inland American lakes. These birds are perfectly hardy, good layers, and very palatable when brought to the table. The plumage of the Cayuga Duck is black, approaching a brown; the feathers of the drake being of a beautiful glossy color, when seen on a bright sunny day looking splendidly. The white color on the neck sets the bird off to good advantage, and with a little pains on the part of breeders might soon be made into a neat, well-defined ring. They can be bred to weigh from six to eight pounds each. The flavor of this bird, in our opinion, is far superior to that of the Rouen or Aylesbury duck, with far greater aptitude to fatten. Its flesh has the juiciness and richness of flavor of the best of our wild ducks.

CAYUGA BLACK DUCK.

THE MUSK OR BRAZILIAN DUCK.

The color of this duck is of a very dark, rich, blue-black prismatic, with every color of which blue is a component, and a white bar is on the wing,

WOOD OR SUMMER DUCK.

some white about the head and neck. The feathers on the back of the male are somewhat fine and plume-like, the legs and feet are dark. This duck is represented as very prolific in a warm climate. The drakes often attain ten pounds weight when well fattened, while the female hardly ever exceeds six pounds. BROWNE says "the Musk duck, in a wild state, is found only in South America." This is a great mistake, for we have often observed them in their wild state in the inlets or bays on Lake Ontario, and as far inland as Cayuga Lake, where they have been shot in great numbers and forwarded to the New-York and Philadelphia markets. It is easily distinguished by a red membrane surrounding the eyes and covering the cheeks.

THE WOOD OR SUMMER DUCK.

This beautiful bird, one of the finest varieties we have, is easily reared and domesticated, and is familiarly known in all parts of the United States. The late M. VASSAR, Esq., of Poughkeepsie, reared them successfully for a number of years, and had them domesticated to such a degree as to permit a person to stroke them on the back with the hand.

THE CRESTED DUCK.

LATHAM, in speaking of the Crested duck, gives the following characteristics of it:—" This inhabitant of the extremity of America is of the size of the wild duck, but is much longer, for it measures twenty-five inches in length; a tuft adorns its head; a straw-yellow, mixed with rusty-colored spots, is spread over the throat and front of the neck; the wing speculum blue beneath, edged with white; the bill, wing and tail are black; irides red, and all the rest of the body ashy-gray."

OTHER BREEDS.

There are a few other pretty varieties of ducks not very common. The White and Black Poland are among the number. They are crested, breed early and are excellent layers. The Labrador also is a rare duck, and highly prized.

GEESE—MANAGEMENT AND DIFFERENT BREEDS

In lieu of anything we can say relative to housing, breeding, rearing, and the general management of geese, we give what the late N. C. BEMENT has written on the subject, with the simple remark that from what we knew of that gentleman when living, we consider his advice orthodox. He says:

GEESE HOUSES, OR PENS.

"In selecting a situation for a goose house or pen, all damp must be avoided; for geese, however much they may like to swim in water, are fond at all times of a clean, dry place to sleep in. It is not good to keep geese with other poultry: for when confined in the poultry-yard they become very quarrelsome, harass and injure the other fowls; therefore it is best to erect low sheds, with nests partitioned off, of suitable size, to accommodate them; and there should never be over eight under one roof; the large ones generally beat the smaller, in which case they should of course be separated, one from the other, by partitions extending out some distance from the nests.

THE NESTS FOR HATCHING

should be made of fine straw, of a circular shape, and so arranged that the eggs cannot fall out when the goose turns them. From thirteen to fifteen will be as many as a large goose can conveniently cover. The ganders remain near when sitting, and seem to watch them as a kind of sentinel; and woe be to man or beast that dares approach them. They seem very anxious to see the young ones, that are to be born, make their appearance.

INCUBATION

lasts from twenty-eight to thirty days, and not two months, as some state, and the goose should have water placed near her, and be well fed as soon as she comes off the nest, that she may not be so long absent as to allow the eggs to cool, which might cause her to abandon her task. After twenty-eight or twenty-nine days' incubation, the goslings begin, but frequently at an interval of from twenty-four to forty-eight hours, to chip the shell. Like turkey chickens, goslings must be taken from under the mother, lest, if feeling the young ones under her, she might perhaps leave the rest of the tardy brood still unhatched. After having separated them from her, they must be kept in a basket, lined with wool and covered with cloth; and when the whole of the eggs are hatched, may be returned to the mother. The male seems to evince the same solicitude for the young as the mother, and will lead and take equal care of them. We once had a gander of the Chinese variety that actually took a brood of goslings from under a common goose, and brought them up with equal care.

ON THE SECOND DAY AFTER THEY ARE HATCHED

they may be let out after the dew is off, if the weather is warm, but care must be taken not to expose them to the scorching rays of the sun, which might kill them. All authors seem to agree on the proper food to be given them, which is coarse barley meal, bruised oats, bran, crumbs of bread soaked in milk or curdled milk, lettuce leaves chopped fine, or crusts of bread boiled in milk. In this country Indian meal moistened with water is generally given; but in our experience we have found it too laxative, and to counteract the effect we have moistened it with boiled milk, and occasionally added chives chopped fine. It is our opinion, however, that more goslings are killed by over-feeding than by starving. A person who is curious in these affairs informed us that he had been most successful when he let the goslings shift for themselves, if the pasture was good. We tried a brood that way and succeeded well. Grass seems to be their natural food, and by following nature in all cases with animals, and more especially with fowls, we have generally succeeded best.

AFTER THEY ARE THREE OR FOUR WEEKS OLD

they may be turned out in a field or lane containing water. If their range is extensive they must be looked after, as the goose is apt to drag the goslings until they become cramped or tired, some of them squatting down and remaining at evening, and are seen no more. After the goslings are pretty well feathered they are too large to be brooded under the mother's wings, and will sleep in groups by her side, and must be supplied with good and renewed straw to sit on, which will be converted into excellent manure. Being now able to frequent the pond and range the common at large, the young geese will obtain their own living; and if favorably situated, nothing more need be allowed them excepting the vegetable produce of the garden. We have, however, found it a good practice to feed a moderate quantity of solid food to the young and store geese, by which means they are kept in a growing and fleshy state, and attain a larger size; the young ones are also forward and valuable for breeding stock. Besides, feeding them, especially in the evening, on their return, attaches them to their home.

DISEASES TO WHICH THEY ARE SUBJECT.

"'Prevention is better than cure;' so says the proverb. Colds and fogs are extremely against geese; therefore, when young, care should be taken not to let them out but in fair weather, when they can go to their food without a leader. They are particularly subject to two diseases; the first a looseness, or diarrhea, for which MAIN recommends hot wine in which the parings of quinces, acorns, or juniper berries are boiled. The second is like

CANADA OR AMERICAN WILD GEESE.

a giddiness, which makes them turn round for some time; they then fall down and die, if they are not relieved in time. The remedy recommended by MAIN is to bleed the bird with a pin or needle, by piercing a rather prominent vein situated under the skin which separates the claws. Another scourge to goslings are little insects which get into their ears and nostrils, which fatigue and exhaust them; they then walk with their wings hanging down, and shaking their heads The relief proposed is to give them, on their return from the fields, some corn at the bottom of a vessel full of clear water; in order to eat it, they are obliged to plunge their heads in the water, which compels the insects to fly and leave their prey.

FOOD AND FATTENING.

" 'It is the same with the goose,' says MAIN, 'as with every other bird that is fattened up; that moment must be laid hold of, when they come to a complete plumpness, or they soon get lean and die if they are not killed.' Meal and skimmed milk will soon do the business; after ranging in the grain stubbles but little else will be required. These are called '*green geese*,' and are most esteemed by the epicure; they will then be about six weeks old, tender and fine. The writer of the article on poultry in *Baxter's Library of Agriculture* recommends steamed potatoes, with four quarts of ground buckwheat or oats to the bushel, mashed up with the potatoes, and given warm. This, it is said, will render geese, cooped in a dark place, fat enough in three weeks. The French method of fattening is detailed very copiously by M. PARMENTIER. 'The whole process,' says he, 'consists in plucking the feathers from under the belly; in giving them abundance of food and drink, and in cooping them up more closely than is practiced with common fowls; cleanliness and quiet being, above all, indispensable. The best time is in the month of November, or when the cold weather begins to set in. When there are but a few geese to fatten, they are put in a cask, in which holes have been bored, and through which they thrust their heads to get their food; but as this bird is voracious, and as with it hunger is stronger than love of liberty, it is easily fattened, provided they are abundantly supplied with the wherewithal to swallow.'

"The Romans considered the liver of the goose a great dainty, and to increase its size they fed them sixteen days on a paste of Turkey figs, stamped and beaten up with cream; their livers would thus be brought to table, each weighing three or four pounds. Equal parts of the meal of oats, rye, and peas, mixed with skimmed milk, form an excellent feeding article for geese and ducks. The grand object of preparing, not geese only, but all kinds of poultry for market in as short a time as possible, is effected solely by paying unremitting attention to their wants; in keeping them thoroughly clean, in supplying them with proper food (dry, soft, and green,) water, exercise, ground, etc. They should be fed three times a day; and it is truly astonishing how soon they acquire a knowledge of the time."

THE AMERICAN WILD GOOSE.

THERE seems to be a great diversity of opinion, among writers on poultry, relative to the domestic or common goose of America, many contending that they derive their parentage from the " Canada Wild Goose," so-called in Europe, while it is said by eminent ornithologists that the American Wild Goose is identical with the Canada, and that the latter derives its name from the former breed. CUVIER claims, however, that the American wild goose, so-called, is identical with the swan family and cannot be well separated from the true swans. But they show

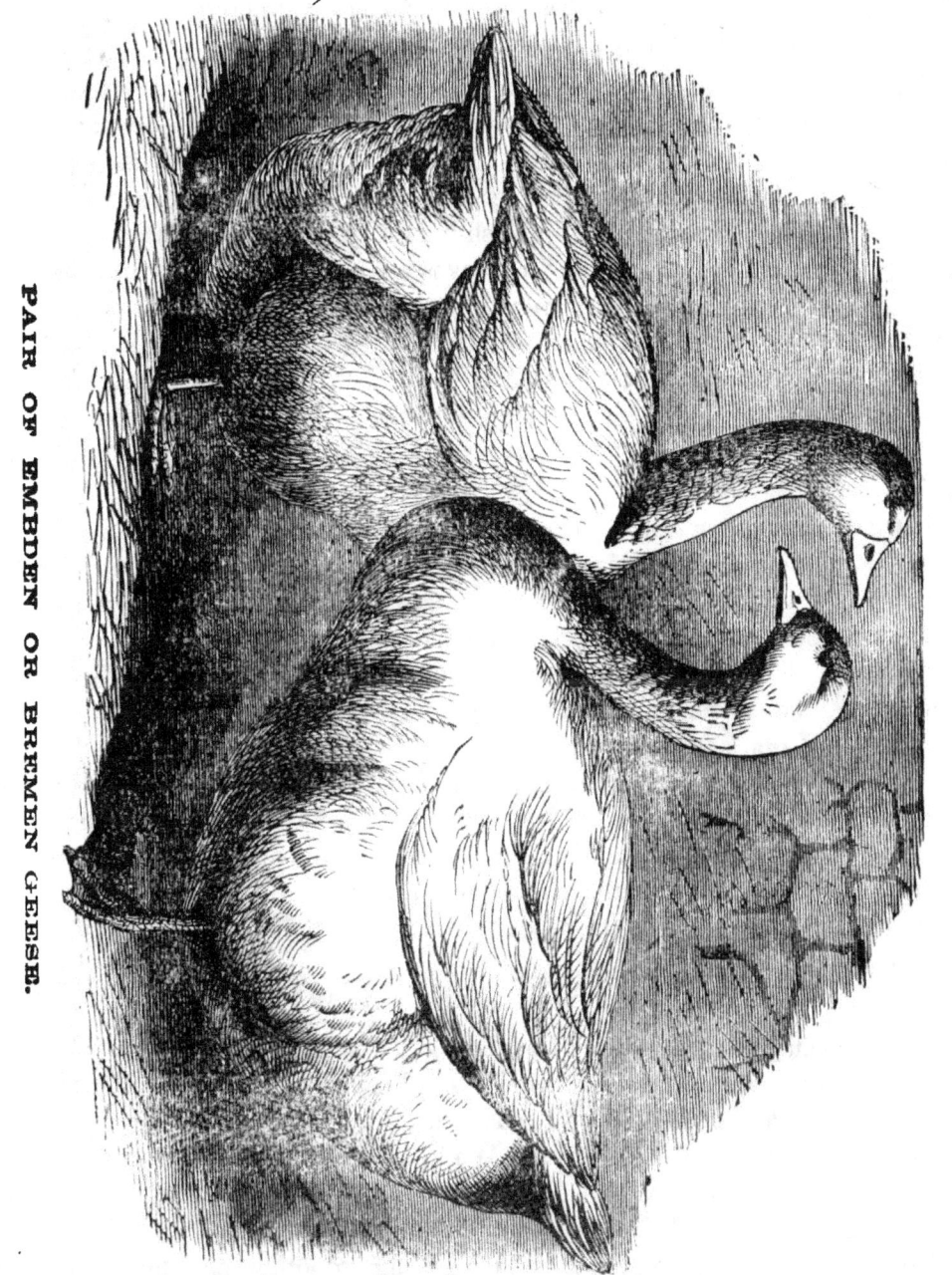

PAIR OF EMBDEN OR BREMEN GEESE.

much more disposition for domestication than the swan, and can certainly be maintained, perfectly healthy, with more limited facilities for bathing than any of the swan family. AUDUBON kept some of the American wild geese three years; yet the old birds did not show any inclination to breed during their confinement; while their young, which were captured with them, commenced breeding the second year. He states their period of incubation to be only twenty-eight days, which is a much shorter period than a person would naturally suppose. In a domestic or confined state they do not breed, as a general thing, until they are at least two years old, while in a wild state they breed when they are from fifteen to sixteen months old.

The American wild goose is undoubtedly one of the most beautiful birds of the feathered tribe, universally known over the whole broad extent of our country, and their regular migrations are a sure signal of approaching winter or the return of spring time. The head, two-thirds of the neck, the larger quills, the rump and tail are jet black; the back and wings are brown, the edges of the wings being a lightish-brown; the under plumage and base of the neck are a brownish-gray; the eyes are encircled with white feathers, while a kidney-shaped cravat of white feathers forms a conspicuous mark on the throat; the upper and under tail coverts are pure white, bill and feet black; while its delicate and swan-like neck gives this bird a majestic and beautiful appearance. Their autumnal flight lasts from the middle of August to the middle of November, and the vernal flight from the middle of April to the middle of May.

WILSON says that, "except in calm weather, the flocks of American wild geese rarely sleep on the water, generally preferring to roost all night in the marshes. When the shallow bays are frozen over, they seek the mouths of inlets near the sea, occasionally visiting the air or breathing holes in the ice; but these bays are seldom so completely frozen as to prevent their feeding on the bars at the entrance." A friend of ours, residing in Chenango county, several years ago, shot and wounded a large gander of this species, clipped his wings, (one of which, we think, was broken,) and placed him with his flock of common gray geese, which paired, and from which he bred a beautiful cross-breed, which were quite a novelty to look upon. The gander seemed to be perfectly contented with his new-found mates, and did not, after a short time, evince any disposition to be freed from the bounden fetters of domestication.

EMBDEN OR BREMEN GEESE.

These beautiful aquatic birds were first introduced into this country in 1821 by Col. SAMUEL JAQUES of Boston, Mass. They were imported direct from Holland; but the appellation of Embden is said to have been obtained from the town of that name in Hanover. We have seen, on several occasions, fine specimens of these geese at the New-York State and other fairs. The flesh of these geese is very different from that of our domestic variety, for it does not partake of that dry character which belongs to other and more common kinds, but is as tender and juicy when brought to the table as that of our wild fowls, and is less liable to shrink in the process of cooking. Epicures aver that the flesh of these geese is not inferior to that of the Canvas-back duck. These fowls are often bred to weigh from seventeen to twenty pounds and upwards. The young are easily reared, with very little care, in almost any section of country. They have been known to weigh, at eight months old, from twelve to sixteen pounds when dressed for the table.

They are the most beautiful of all geese, and, excepting the Toulouse, the largest. Indeed, the rivalry between the two breeds is so close that many contend that the palm of size as well as beauty belongs to the Embden.

Mr. HEWITT, an English writer who favors this variety, says:—"The Embden goose has prominent blue eyes, is remarkably strong in the neck, and the feathers, from near the shoulder to the head, are far more curled than is seen in other birds. The plumage is pure white throughout; bill flesh color, and legs orange. One of their great advantages is this:—That all the feathers being perfectly white, their value, where many are kept, is far greater in the market than is ever the case with colored or mixed feathers. The quality of the flesh is about equal with the Toulouse; but the Embden is the earlier layer, and frequently rears two broods in one season, the young ones proving as hardy as any with which I am acquainted."

THE TOULOUSE GOOSE.

The Toulouse goose is said to have originated in France, and is distinguished from the common gray goose by its colors being darker and more intense, by the bright orange hue of the bill, legs and the orbit around the eye, as also by the singularly early development of the abdominal pouch. The Earl of Derby first introduced this breed in England from the south of France, and, like the Embden, they attain to great size. They are good layers, and their flesh is tender and well flavored. DIXON, in describing the goose, says:—"The head should be depressed, and of a more elongated form than in the common goose; bill three inches in length by two inches in depth at the base; in color a clear orange-vermilion, the nail at its extremity being white, irides dark brown; orbit large, and of the same color as the bill. The plumage of head and neck ash-gray, the latter showing 'the curl' in a very marked manner. Throat a light tint of gray; breast, back, and thighs dark grayish-brown, with a margin of white, more or less distinct, on each feather. Greater wing coverts brown; lesser wing coverts a light gray. Primary wing-feathers, of which the second is the longest, ash-gray, becoming very dark, rich brown at their extremities, the shaft being a clear white; secondaries and tertials dark leaden-brown; scapulars the same, with a narrow light edge. Under part of the body white; tail-coverts white; tail-feathers brown, with broad white band at the extremity. Legs and feet reddish-yellow; claws dusky. The wings, when folded, about half an inch shorter than the tail. The orbit, in both its form and color, the general tone of plumage, the color of the bill and legs, the particular light marking of the lesser wing coverts, and the wings, which fall short of the tail, are points of resemblance between the Toulouse and the Gray-leg goose."

THE AFRICAN GOOSE.

Africa, or, perhaps, some of the southern countries of that old continent, seems to be the native abode of this goose. They are the largest of the goose tribe, and often weigh twenty-five pounds and upwards. Although LINNAEUS, in his description, has termed them Siberian geese, they are not indigenous in Siberia, but have been carried hither and multiplied in a state of

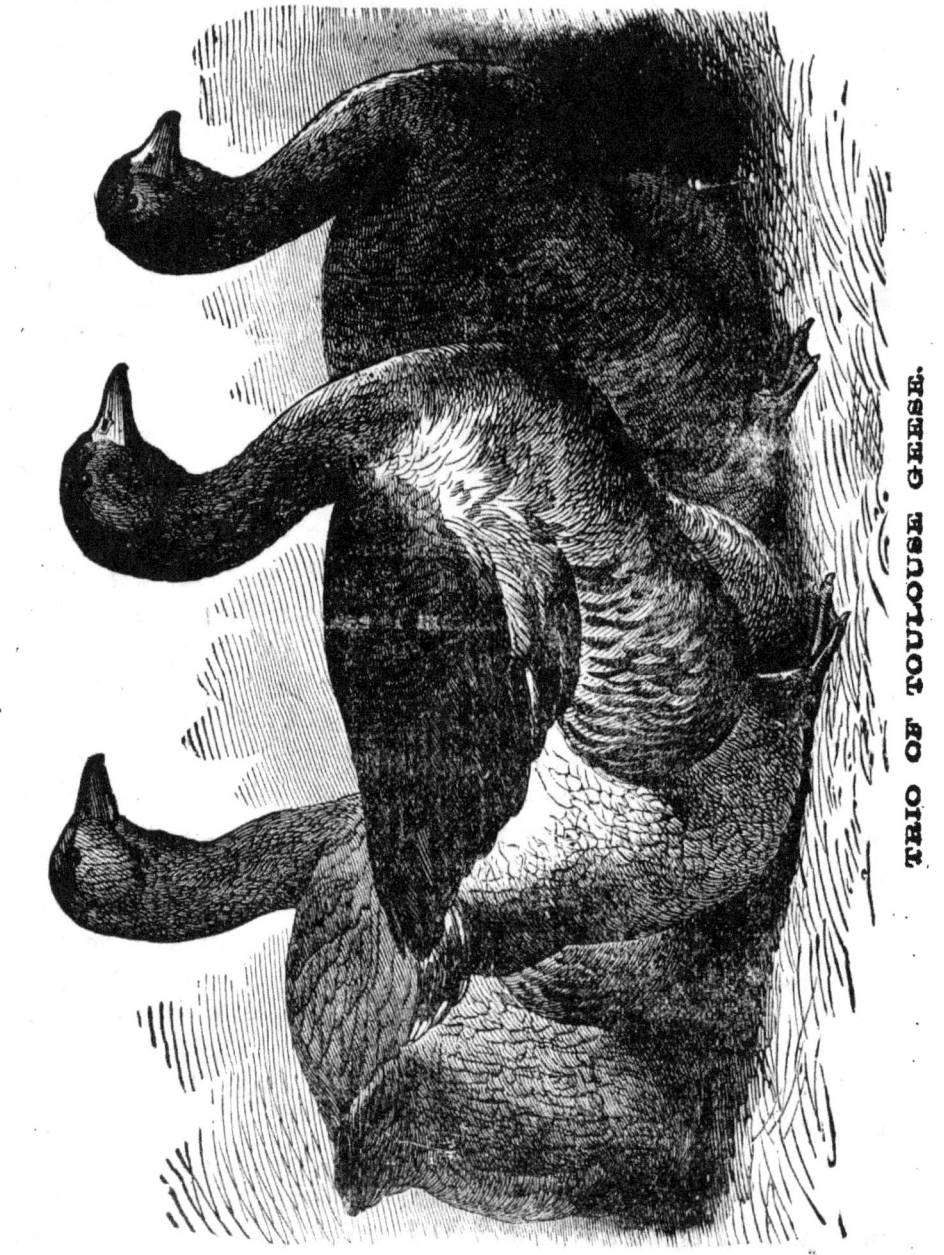

TRIO OF TOULOUSE GEESE.

domestication, as in Germany and Sweden. This bird carries its head high as its walks, and its fine carriage and great bulk give it a noble air. The bill is armed at the edges with a small indentation, the head and the top of the neck are brown, deeper on the upper side than on the under; on the origin of the bill there rises a round and fleshy tubercle of a vermilion color; under the throat, also, there hangs a sort of fleshy membrane, which is firm and hard.

THE EGYPTIAN GOOSE

is bred in this country to a certain extent. It is a beautiful and stately bird, and is much valued for its gorgeous mantle of golden hues. It is also very prolific, bringing off, usually, three broods a year, from eight to twelve each time. Their weight is about eight pounds each. Their markings are striking and beautiful, being dark-red round the eyes; red ring round the neck; bill white; neck and breast light fawn-gray; a maroon star on the breast; belly red and gray; half of the wing-feathers rich black, the other

WHITE CHINESE GOOSE.

part of them pure white; black bar running across the center; back light-red, growing dark-red toward the tail; the tail a deep black; carriage upright and stately.

THE WHITE CHINESE GOOSE.

This variety of aquatic fowl was introduced into England some years since by ALFRED WHITIKAR, and brought to this country by JOHN GILES of Connecticut. Mr. WHITIKAR gives the following description of it:—" The White China Goose is of a spotless, pure white, more swan-like than the brown variety, with a bright orange-colored bill, and a large orange-colored knob at its base. It is a particularly beautiful bird, either in or out of the water, its neck long, slender, and gracefully arched when swimming. It breeds three or four times in a season, and its period of incubation extends to five weeks. They are prolific layers, but their eggs are small for the size of the bird, being not more than half the size of those of the common goose.

The spring goslings are easily reared, and are a fair average quality for the table. The disparity in size between the sexes is considerable, often amounting to over one-third of their relative weights. Its color, as its name indicates, is a pure, spotless white, which, contrasted with its yellow or orange-colored bill and legs, gives quite a pleasing effect, and it certainly deserves to rank in the first class of ornamental poultry."

THE BARNACLE GOOSE.

The Barnacle breeds in Iceland, Greenland, and the north of Russia and of Asia. It is of handsome form, standing high on its limbs. The flesh is excellent, and they weigh about eight pounds a pair. The bill is small and black, with a reddish streak on each side; the cheeks and throat, with the exception of a black line from the eye to the beak, white; head, neck, and shoulders black: under plumage marbled with blue, gray, black, and white; tail black; under parts white; legs dusky. Although the Barnacle is shy and cautious in a wild state, yet when brought under a state of domestication it is as tame as any of the goose tribe.

THE BRANT GOOSE.

This and the Barnacle goose are the smallest of their tribe yet introduced to our aquatic aviaries; both being less in size than some ducks. The Brant is considered one of our most savory birds. In its transit from its breeding-places near the Arctic sea, it appears in great numbers on the coast of New-York in the first and second week in October, and continues passing on to the south until December. Some few have been observed to remain all winter. They are again seen with us in April and May, on their way north, when they are in the best condition. "Immense numbers of Brant geese," says Mr. St. John, "float with every tide into the bays formed by the bar. As the tide recedes, they land on the grass, and feed in close packed flocks. On the land, they are light, active birds, walking quickly, and with a graceful carriage. On any alarm, before rising, they run together as close as they can; thus affording a good chance to the sportsman, who may be concealed near enough, of making his shot tell among their heads and necks."

DISEASES OF POULTRY.

SYMPTOMS, CARE, TREATMENT, PREVENTIVES, REMEDIES, ETC.

IN the climate of this country there is no need of having any diseases among our domestic poultry if proper care and judgment in the treatment of the same were manifested on the part of the breeder. We have given in this connection a series of diseases that are known to infest poultry yards not properly cared for, with preventives and remedies for the same, in the hope, that should occasion require, benefit may be derived therefrom.

Apoplexy.—Fowls are attacked with this disease when apparently in the most robust health—suddenly fall down, die, or are found without sensation or the power of locomotion. Bleeding is recommended for the disease; take a sharp-pointed pen-knife and open one of the largest veins under the wing in a longitudinal direction, by pressing the thumb on the vein at any point between the opening and the body, the blood will flow freely and relieve the fowl at once. Stimulating food should not be given to fowls liable to this disease.

Black Rot.—The symptoms of this disease are blackening of the comb, resembling mortification; swelling of the legs and feet, and general wasting of the system. It can only be cured in the earlier stages by frequent doses of castor-oil, to keep up purging; at the same time giving freely strong ale or other stimulants, with warm and nourishing food.

Catarrh in Chickens.—The symptoms of this disease are not dissimilar to those in the human subject, being a watery or slimy discharge of mucus from the nostrils, swelling of the eyelids, and, in extreme cases, the sides of the face are swollen. The cause of the disease is somewhat similar to that of roup. It is said if this disease is not promptly attended to it frequently terminates in roup. Food, consisting of boiled mashed potatoes, well dusted with black pepper, is good. Pills, made the size of a large pea, of mashed potatoes, with cayenne pepper placed in the center, and given to them every other day, at feeding time, for a few days, will insure a radical cure, and give the fowls a good appetite. Dr. BENNET claims that the following will also prove efficacious—it never having been known to fail:—Take finely pulverized, fresh burnt charcoal, and new yeast, of each three parts; flour, one part; pulverized sulphur, two parts; water, quantity sufficient to mix well, and make into boluses of the size of a hazelnut, and give one three times a day." Cleanliness he claims to be essential in all cases, and frequent bathing of the eyes and nostrils of the fowls with warm milk and water.

Chicken Cholera.—A correspondent of the Department of Agriculture, writing from Iowa, says:—" My chickens have been dying with cholera for the last two years,—even turkeys have died of the same disease. When I notice the fowls begin to droop and look sleepy, I give them three or four tablespoonfuls of strong alum water, and repeat the same the next day. I also mix their feed with strong alum water, feeding twice a day for two or three days—afterwards once a week. Since commencing this practice I have not lost any." Another good cure is to give as feed cooked Indian meal, red pepper, gunpowder and turpentine, mixed together. Put in a day's feed, for a dozen fowls, a tablespoonful each of red pepper, gunpowder and turpentine, well mixed through the meal. Give them this food every other day for a week or so, and it will in most cases effect a cure. Another remedy for this disease is to one gallon of water add one ounce of bi-sulphate of soda; set it where the fowls can drink it. As a preventive it is necessary to have the roosting place for the fowls dry and clean; the place where they roost should be cleaned as often as once a week, and sprinkled with lime or wood ashes. Feed with dry feed.

Crop Bound Fowls.—If the crop feels hard and stone-like to the touch, it will be necessary to make an incision with a sharp knife through the skin and upper part of the crop and loosen the unpacked mass by some blunt-pointed instrument, and remove it. The incision, if small, may be left, but if large, a stitch or two is advisable. The birds should then be fed warm, soft food for two or three days,—such as mush and potatoes, well mixed with cayenne pepper and gentian; give them plenty of exercise in the open air, and they will rapidly recover from the disease.

Curling in of the Toes of Fowls.—Large fowls, such as Brahma or Cochin China, and others, are subject to corns in the fleshy part of the foot. These should be opened, the corn extracted, and the wound dressed with a little Venice turpentine, spread on soft cotton or lint, and the foot bound up.

Diphtheria—Is a disease which originates mainly from improper care and sudden changes of weather and variations of temperature. It affects fowls of all ages; is either acute or chronic, sometimes beginning suddenly, at others gradually, and seems a kind of lingering consumptive disease. It is also occasioned by improper and damp coops and roosts. Fowls, to escape the roup, catarrh, pip, gapes and similar diseases, should be fed on wholesome food and placed in dry, well ventilated coops—cleanliness proving a great assistance to health. It makes its appearance in a way similar to the croup in the human being. It fills up the windpipe at its opening with a sort of white ulcerous substance, and continues to form and spread over the entire tongue and mouth, occasionally causing the fowl to cough, raise its head, and open its mouth to breathe. The smell from it is very offensive, and unless the bird is relieved it pines away and dies. The best cure for this disease that we have heard of being used with any degree of success is nitrate of silver and powdered borax. Remove the ulcers as much as possible, and apply the nitrate of silver with a feather. Powdered borax can be applied in the same manner by wetting the feather, dipping it in the powder, and swabbing the throat. A little chloride of potassium dissolved in the water which is given the fowls to drink, may possibly avert the disease—say one-quarter ounce to a half gallon of water.

Dust Baths.—Fowls in confinement need a dust bath, *i. e.*, a box of mixed ashes and earth to wallow in. An ordinary soap box will do, filled two-thirds full of dry earth and wood or coal ashes. Wood ashes is preferable when it can be obtained. There is no better preventive of lice than this; and the fowls enjoy it hugely.

Dysentery in Fowls.—Fowls attacked with this disease should be given chalk, mixed with boiled rice and milk; a little alum dissolved in their water, so as to make it a little rough, will be useful. The food should be dry grain; no food of a laxative tendency should be given them.

Egg-Bound.—To relieve a hen that is egg-bound, take a common tail feather of the hen and strip it until near the tip, and then dip it in sweet oil, and let it remain until it becomes thoroughly saturated, then pass the feather up the egg-passage till it meets the egg, which you will find will relieve the hen at once, and enable her to proceed with her duties; if she experience any further difficulty, repeat the operation, getting the feather well filled with oil whenever you make an application. Do not attempt to help nature, in the way of pressure, for in that case the egg may become broken and prove fatal to the hen. After you have made the application, as directed, let nature take its course, and all will be right.

Enlargement of Liver and Gall.—This frequently occurs in over-fed fowls, or in consequence of feeding unnatural or over-stimulating food. For a remedy, feed soft cooked food, so as to make as little call upon the digestive organs as possible; give a grain of calomel every other day for a few days, and remove the bird to dry, warm quarters.

Fowls Eating their Feathers.—To prevent fowls eating their feathers give them animal food, such as fresh meat, two or three times a week, burnt bones, oyster shells, charcoal, together with good clean water and hennery. If this does not produce the desired effect, wring their necks, for nothing else will prove a cure.

Frost-Bitten Combs.—Frost-bitten combs can be cured by making a thorough appplication of glycerine three times a day.

Gapes in Fowls—Is no new disease, but one with which every poultry breeder and fancier should make himself as familiar with as "household words;" for all domestic birds are liable to it, more particularly all young fowls, if not properly guarded against. This disease is most destructive in the excessively warm weather of July and August. It is caused by ill-ventilated and unclean coops, together with the unwholesome, sour food and putrid or impure water, too often given to young fowls; more particularly is this the case with young turkeys. It should be borne in mind, also, that the "gapes" is an epidemic disease, and when it once make its appearance in a flock of young fowls, those affected with it should at once be removed from the coop; for it is well understood that "an ounce of prevention is better than a pound of cure." The gapes is said to be caused by a sort of internal worm infesting the wind pipe; in some instances it has been so observed, but it is by no means a sure criterion with all the disorders accompanied with the gaping of fowls. The general symptoms of the disease, and those most noticeable, are the continual gaping, coughing, dullness, inactivity, loss of appetite and sneezing of the fowls attacked. Mr. MOWBRAY, an eminent English breeder, says the "disease first shows itself when the chicken or turkey is between three and four months old, and not often after." On the contrary, we have seen the disease in its worst form show itself in young turkeys and chicks from four weeks to six months of age; therefore, there is no more certainty of fowls being rid of the disease at four weeks old than they are at six months old. There are several modes for the treatment of turkeys for this disease; the one which has proved the most successful in cases which we have treated, is as follows:—Take a small quill feather, stripping the vane, except half an inch from the extremity, of the feathers; this should be dipped in spirits of turpentine, and the diseased turkey or chicken, as the case may be, being held, the feather so prepared, is passed down through the small opening of the wind-pipe, which is readily seen at the base of the tongue, and giving it one or two turns, will generally bring up and destroy the worms. The turpentine at once kills the worms, and its application excites a fit of coughing, during which those that are not drawn out by the feather are expelled by the coughing. After this process being used, the young turkeys should be kept for several days in a dry coop, and not be allowed to wander in damp, swampy places, or wet grass. Their feed should be either cooked corn meal or cracked wheat, which is better, soaked in turpentine, given every morning, and the remainder of the day they should be fed with boiled whey or sour milk, well sprinkled with black pepper; they should also have plenty of clean, fresh water in the coop. Crushed corn soaked in alum water is also said to be a good remedy for gapes. BEMENT, in the *American Poulterer's Companion*, recommends shutting up the turkeys or chickens in a box, with some shavings dipped in spirits of turpentine, when the vapor arising from the extended surface, produces, in most cases, a cure. He also recommends creosote, used in the same manner, which will produce a like result. We know nothing of the efficacy of these remedies, never having tried them; but we do know that spirits of turpentine will not harm fowls if it does not do them any good. The remedies are certainly simple, and no doubt well worthy of trial.

Gout or Swelled Legs.—It is recommended for this disease to rub the leg of the fowl affected with fresh grease of any kind once a day for a week, when a cure will be effected. Another remedy is to give a grain of calomel at night, and three drops of wine of colchicum twice a day, care being taken as to warmth, diet, etc., of the fowl.

Leg Weakness.—This disease occurs in highly-fed, fast-growing chickens. Give them animal food once a day, and in *warm* weather dip the legs for a few minutes daily in cold water; also give them every day three or four grains of ammonio-citrate of iron dissolved in water and mixed with meal-feed. Keep them from the wet grass.

Pip.—The pip is occasioned by the forming of a dry, horny scale upon the tongue — the beak becomes yellow at the base, the plumage becomes ruffled, the bird mopes and pines, the appetite gradually declines to extinction, and at length it dies, completely worn out by fever and starvation. Give the bird, three times a day, for a week or so, two or three grains of black pepper in fresh butter, which will effect a cure.

Rheumatism.—This disease is caused by exposure in cold, damp and wet henneries. It may be prevented by placing the fowls in warm and dry locations, free from chilling rains and cold, bleak winds. Feed cooked Indian meal and potatoes, made into a mash, mixed with ale, blood warm, twice a day. Local applications are useless.

Roup.—The symptoms of this disease are somewhat similar to those of catarrh. The bird has a frothy substance in the inner corner of the eye; the lids swell, and in severe cases the eye-ball is entirely concealed, and the fowl, unable to see or feed, suffers from great depression, and sinks rapidly; the fœtid smell being unbearable. In aggravated cases the following will be found beneficial:—Powdered sulphate of iron, half a drachm; capsicum powder, one drachm; extract of licorice, half an ounce; make into thirty pills; give one at a time three times a day for three days; then take half an ounce of sulphate of iron, and one ounce of cayenne pepper in fine powder. Mix carefully a teaspoonful of these powders with butter and divide into ten parts; give one part twice a day. Wash the head, eyes, and inside of the mouth and nostrils with vinegar; it is *very* cleansing and beneficial. Another remedy for this disease, one which rarely fails to cure, is to take nitric acid, strip a feather to within half or three-fourths of the end, dip the feather into the acid, and thrust it into the nostril of the sick bird, giving it a twist while in. Repeat this twice or three times a day, removing the burnt scab *before* applying the acid. It is rarely necessary to make a fourth application, and very frequently one is sufficient. Mrs. ARBUTHNOT's remedy is confinement alone in a warm, dry place; a tablespoonful of castor oil every morning for a week; feed with soft food only, mixed with ale and chopped vegetables. In all cases where the bird is attacked with this disease it should be separated at once from the coop, and placed in a good dry, warm location, and not allowed to mix with other fowls on any consideration.

Scurvy Legs.—Fowls that show any symptoms of this disease should at once be removed from the pen and placed in warm, dry quarters. Give them wholesome and animal food as often as once a day; wash the legs with a *weak* solution of sugar of lead in the morning, and anoint them with clean lard, mixed with ointment of creosote in the evening, just before going to roost. Don't, on any consideration, allow the fowls to be exposed to drenching rains or roam in wet or even damp grass; but keep them warm and as quiet as possible until the disease disappears, which, if proper care is taken of them, will result in from eight to ten days.

To Exterminate Lice.—There are almost as many remedies for ridding the hennery of lice as there are breeds of fowls. We will in this connection give a remedy which we have tried with success — one answering all purposes desired. We will guarantee, if the directions are followed, it will exterminate both the common hen louse and the minute hen spider, (the last named being the worse of the two). Take all the hay from the nest and burn it. Drive all the hens out. Get an iron pot or vessel of any kind, put it in the center of the house; shut the house as tight as it can be; put in the pot a pound of roll brimstone. Heat a piece of iron as large as a man's fist red hot and put in the pot with the brimstone. Keep the house shut close two hours, then open and ventilate. Sweep and dust out the house thoroughly. Dissolve one pound of potash in one quart of hot water. With an old paint brush paint or wash every part of the house, inside and out, roosts, nests and every place that can be reached with the solution. Get, now, a quart of kerosene oil and go through the same operation, painting the whole inside of the house, saturating the roosts well with it. There will not be a louse left when these directions have been followed. It is some work to do it, but it will pay. Put fresh hay in the nests and let the hens in. When they go on the nests to lay, as soon as the nest is warm, if there are any lice on them the latter will leave. They will be seen crawling around the front of the nest boxes; but their lives are short; they cannot endure this remedy and live.

Vertigo.—Fowls affected with this disease, BEMENT says, may be observed to run round in a circle, or to flutter about with but partial control over their muscular actions. The affection is one evidently caused by undue determination of blood to the head, and is dependent on a full-blooded state of the system. Holding the head of the fowl under a stream of cold water for a short time immediately arrests the disease; and a dose of any aperient, such as calomel, jalap, or castor oil, removes the tendency to the complaint.

White Comb—Makes its appearance in the form of small white spots on one or both sides of the comb, which are so thickly clustered together as to be mistaken for a sprinkling of meal or other white powder. It seems to be of a scorbutic or leprous nature. The disease spreads itself down the neck of the fowl, both in front and back, and takes off all the feathers as far as it goes, leaving only the stumps. TEGETMEIER recommends stimulating, wholesome food, say oatmeal and water, with a supply of green vegetables, and the administration of some alterative medicine, as flour of sulphur, ten grains, and calomel, one grain, given every other night; and anoint the comb with fresh lard. It can be successfully cured by using cocoa-nut oil, powdered turmeric and sulphur, made into an ointment, and anointing the part affected three or four times a day, and an occasional dose of six grains of jalap. The proportions are about a quarter of an ounce turmeric powder to one ounce of cocoa-nut oil, and a third of an ounce sulphur.

Tonic for Poultry.—Mr. MILLS, an apothecary of considerable note in Bourges, France, in the *Journal d'Agriculture Pratique*, recommends the following prescription — one which he avers he has used successfully — as an invaluable tonic for debilitated birds, especially in the mortality which is apt to prevail when "shooting the red." He says:—"Take cassia bark in fine powder, three parts; ginger, ten parts; gentian, one part; anise seed, one part; carbonate of iron, five parts. Mix thoroughly by sifting. A teaspoonful of the powder should be mingled with the dough for twenty young turkeys each morning and evening. It is of the greatest importance to begin the treatment a fortnight before the appearance of the red, and to continue it two or three weeks after."

Molting Fowls should have a few nails placed in the water furnished for their use. The rust occasioned by nails renders fowls less liable to disease.

Tansy is almost a certain preventive of lice upon setting hens. Gather it green, and line the nest, at the time of setting the hens.

FRACTURES OF THE BONES.

In regard to this matter TEGETMEIER says, that "fractures of the bones of the body are less likely to occur in birds than in other animals, inasmuch as the framework is more completely united together, and is protected from injury by the feathers. In cases where fracture of the ribs or other bones may be suspected, there would be great difficulty in determining the nature of the injury, and I do not think anything more could be done than keeping the bird quiet until recovery. In cases of broken wings, the quill feathers would prevent any recourse being had to the ordinary method of bandaging. The plan I have pursued is, to tie, carefully, the ends of some of the quills together in their natural position, with the wing closed; this prevents motion of the broken ends of the bones; and by keeping the bird in an empty place, where there are no perches for it to attempt to fly upon, every chance of recovery is afforded. Fracture of the fleshy part of the leg would be less manageable, and I can hardly recommend any bandaging that would be readily applied. The most common fracture in fowls is that of the tarsus, or scaly part of the leg. This is usually treated by wrapping a slip of rag round the injured limb, and tying it with thread — a very imperfect plan, as motion of the broken bones is not prevented, and which is therefore frequently unsuccessful in its results. I always employ a modification of what is known to surgeons as a gum splint. The white of an egg is well beaten up with a fork, and spread upon a strip of thick, soft brown paper, as wide as can be smoothly wrapped around the broken limb. The fowl is held by an assistant, the leg slightly stretched, so as to bring the ends of the bones in a straight line, the moistened paper wrapped smoothly round several times, and secured by two or three turns of thread; and, lastly, to prevent the parts being moved before the paper has become dry and stiff, a thin splint of wood, such as is used for lighting pipes, bound with thread on each side; the wood may be removed the following day, as it then adds to the weight. The stiff paper forms a bandage which prevents all motion, and so places the limb in the best possible condition for union to take place."

SULPHUR FOR FOWLS.

We have been advised by a lady friend, who is no novice in rearing poultry, that no one who has not had the experience, can imagine the beneficial effect a little sulphur mixed with the food of fowls and given two or three times a month, will have upon them. Sulphur is good to be given in all cases, and seems to permeate through the system of the fowl, promoting health and preventing disease. Mix, with the feed intended for a dozen fowls, about half an ounce of pulverized sulphur.

POULTRY HOUSES, YARDS AND RUNS.

A PLAN OF A POULTRY HOUSE THAT WILL ACCOMMODATE FROM TWENTY TO THIRTY FOWLS.

To those wishing a small hennery or duck house, for the accommodation of from twenty to thirty fowls, we commend the following directions as worthy of consideration:—In the first place, the house should be in a situation that is dry and airy, but not exposed to tempests; the aspect warm,—an eastern or southeastern location is the best,—sheltered, if it may be, by a screen of trees or shrubbery, so that the birds may have the shelter thereof from the summer midday sun, and raw, inclement winds of winter. The house should also be constructed so as to give as much warmth as possible, with a perfect command of ventilation. The floor should be elevated over the general surface, so as to be perfectly dry; the walls close and substantial; the roof air and water-tight; windows should be placed opposite each other to admit of thorough ventilation; but one should be closed at night, even in summer, to prevent through draft during sleeping hours. The windows should be latticed to prevent the fowls passing out or in. The roosting perches should commence at about a foot from the ground, and ladder-ways, placed twelve inches or so apart, and rising twelve inches, one above the other, for cocks and hens. Turkeys require eighteen inches rise, and at least two feet apart. The perches to be one and a-half to two inches in diameter, with the angles taken off, but not made smoothly round; nests to be constructed in the end walls. The house for twenty fowls should be between five and six feet long, ten feet deep, from front to rear, seven feet high at the front, and nine or ten feet high at the back. That for turkeys must be seven or eight feet long, and the same depth, hight, etc., of the other houses. That for ducks may be of the same dimensions as the hen house, but requires no perches. A feeding coop may be made in the bottom compartment, two feet wide and two feet high, to suit the large birds; the upper one eighteen inches wide and eighteen inches high, for the smaller ones; the sides and ends to be closely boarded; the front to be done with rounded railing, in which the doors are to be made, also railed, through which to take out and put in the fowls; or the backs may have the doors in them. Along the front the feeding troughs are to be placed. These coops may be placed in a compartment in the same range with the other houses, or one resting against the back of the poultry houses.

A RUSTIC POULTRY HOUSE.

We can scarcely fancy anything more beautifying in a poultry yard than a nice and convenient rustic poultry house, combining convenience with simplicity. In this line we find nothing more suitable than the following, which we take from BEMENT's *Poulterer's Companion*: — For the rustic work, join four pieces of saplings in an oblong shape for sills; confine them to the ground; erect at the middle of each of the two ends a forked post, of suitable hight, in order to make the sides quite steep; join these with a ridge pole; put on any rough or old boards from the apex down to the ground; then cover it with bark, cut in rough pieces, from half to a foot square, laid on and confined in the same manner as ordinary shingles; fix the back end in the same way; and the front can be latticed with little poles with the

A RUSTIC POULTRY HOUSE.

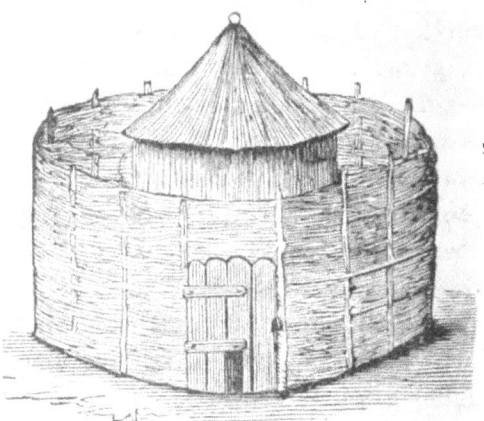

THE POOR MAN'S POULTRY HOUSE.

bark on, arranged diamond fashion, as shown in the engraving. The door can be made in any style of rustic form. The roosts, laying and sitting boxes can be placed inside of the house, in almost any position; either lengthwise or in the rear. From the directions here given, a person can easily build a fancy rustic house of any desired size, and in almost any location in the poultry yard desired. To make the rusticity of the house show off to the best advantage, it should be placed amid shrubbery.

THE POOR MAN'S POULTRY HOUSE.

The plan is a cheap and economical one—such as can be built with very little trouble or expense, combining at the same time a good and convenient poultry yard and house by simply thatching it with straw and brushwood instead of using lumber. The *Rural Farmer's Library* says it is made by forming a circle eighteen or twenty-four feet in diameter, in accordance with the size you wish to build; on the outside of the circle cut a trench, three or four inches wide and deep, and plant poles twelve or eighteen inches into the ground every two feet. These poles should be as thick as a

PERSPECTIVE VIEW OF BROWNE'S POULTRY HOUSE.

man's arm, eight or ten feet high; a space on the south side, between two poles, should be chosen for a doorway. Then take brushwood, six feet long, with the twigs and leaves on, place it against the poles and commence lacing some of the stout and straight twigs round the poles in the trench, alternately twining in and out, similar to basket-work, going the whole round,

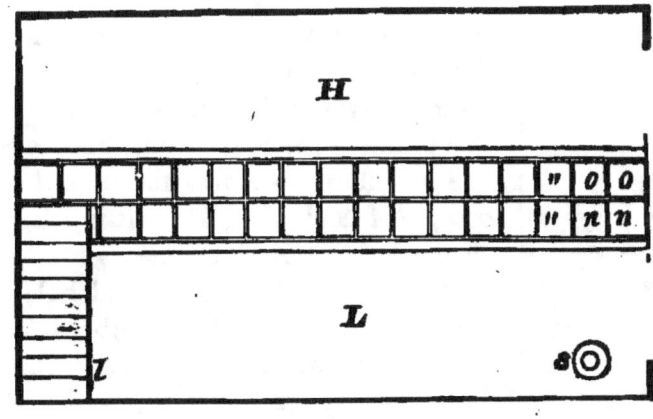

GROUND PLAN.

except the doorway. When eight or ten inches high, stamp it well down, making it tight and firm. Keep on in the same manner until you have got five feet high, then pass the brush over doorway and all, which will make it firmer and stronger, continuing up to eight or ten feet in hight; braid sometimes on one, and then on the other side of the uprights. The house should be placed in the center of this circle. A few stakes, a little more brush, and an armful of straw for thatch or roof, will make this answer; the brush must be woven round the poles in the same manner for the house that it was for the yard. A straw thatch for roof, it is said, will last twenty years, if properly made. It should be formed of good, clean, long straw, and as little broken as possible. Wheat or rye straw is preferable; put it on ten or twelve inches thick; some roofs are made eighteen inches thick. Tie it close and securely with strips of white oak or hickory bark well twisted; but this every one knows how to perform. The roof should have a good pitch, or, in other words, be very steep, so that rain or snow may be quickly thrown off. Doors for this house may be made of boards, and hinges from the sole of an old shoe. The inside of the house may be arranged as desired as regards laying boxes, roosts, etc. The inside of the house might be thatched with straw, as well as with brushwood, which will make it warmer in winter. With the directions here given, and the illustration before him, almost any handy lad upon the farm can build a comfortable hennery and yard.

BROWNE'S POULTRY HOUSE.

From the *American Poultry Yard*, by D. J. BROWNE, we take the following description of a very pretty and convenient poultry-house, of which we give a perspective view:—"A fowl-house," says Mr. BROWNE, "should be dry, well-roofed, and fronting the east or south; and if practicable, in a cold climate, it should be provided with a stove, or some other means for heating, warmth being very conducive to health and laying, though extreme heat has the contrary effect. The dormitory, or roost, should be well ventilated by means of two latticed windows, at opposite ends of the building; and it would be desirable to have one or more apertures through the roof for the escape of foul air. The sitting apartment, also, should be ventilated by means of a large window, in the side of the house, and holes through the ceiling or roof. If kept moderately dark, it will contribute to the quietude of the hens, and thus favor the process of incubation. The sitting room should be provided with boxes or troughs, well supplied with fresh water and proper food for the hens during the hatching period, from which they can partake at all times at will. The laying-room, in winter, should have similar boxes or troughs containing old mortar, broken oyster-shells, soot, brick-dust, gravel and ashes, as well as a liberal supply of proper food and drink. The perches, or roosting poles, should be so arranged that one row

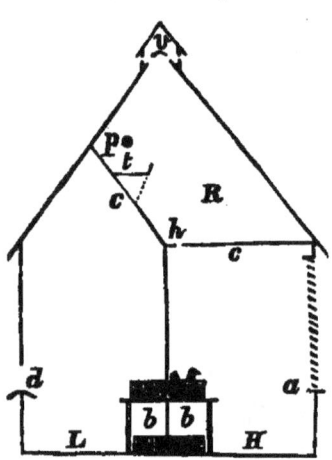

TRANSVERSE OR CROSS SECTION.

of the fowls should not rest directly over another. They should be so constructed as to enable the fowls to ascend and descend by means of ladders or steps, without making much use of their wings; for heavy fowls fly up to their roosts with difficulty, and often injure themselves by descending, as they alight heavily upon the ground. The illustration given represents a hen-house in perspective, twenty feet long, twelve feet wide, and seven feet high to the eaves, with a roof of a seven foot pitch, a chimney-top, a ventilator on the peak, twelve feet in length and one foot or more in hight, and openings in the gable ends for the admission of fresh air. In the easterly end there are two doors, one leading into the laying apartment and loft, and the other into the hatching-room. In the same end there is also a wooden shutter or blind, which may be opened whenever necessary to let air or light into the roost. In the back, or northerly side, there is a large lattice window, three feet above the floor or ground, four by twelve feet, for the purpose of affording fresh air to the sitting hens. In front, or southerly side, there is a large glazed window, four by twelve feet, and another on the southerly side of the roof, of a corresponding size, designed to admit light and heat of the sun in cold weather, to stimulate the laying hens. In the southerly side there are also two small apertures three feet above the ground or floor, for the ingress and egress of the fowls. These openings may be provided with sliding shutters, as well as 'lighting boards,' inside and out, and may be guarded by sheets of tin, nailed on below them, to prevent the intrusion of rats, weasels, or skunks. The building may be constructed of wood or other materials, and in such style or order of architecture as may suit one's taste, only preserving the internal arrangements and proportions in reference to breadth and hight. As a general rule, as regards the length of a building, each hen, irrespective of the cocks, may be allowed a foot. In the ground plan, L denotes the laying apartment; H the hatching-room, six by twenty feet; n, n, etc., nest-boxes for laying, fourteen by fourteen inches, and ten inches deep; $o, o.$ etc., nest-boxes for sitting hens, of the same size; l, a ladder or steps leading into the loft; and S, a stove for warming the apartment, if desirable, when the weather is cold. The transverse or cross section shows the building from the bottom to the top, with the internal arrangements; L denotes the laying apartment, and H the hatching-room, divided in the middle by a partition; n, the nest-boxes resting on tables, three or four feet above the floor or ground; b, b, boxes or troughs containing water, grain, brick-dust, sand, ground oyster shells, or the materials for the convenience of the fowls; d, an aperture or door three feet above the ground or floor, for the ingress and egress of the fowls; a, a lattice window, three feet above the the floor or ground, for the admission of fresh air to the sitting hens; R, the roosting place, or loft, shut off from the laying and sitting apartments by the ceilings, c, c; h, a hole or opening in the ceiling for the escape of the air below into the loft; v, the ventilator at the peak of the roof; P, the roosting-pole, or perch; t, a trough, or bed, for retaining the droppings or dung."

A MODEL HENNERY.

Among the multiplicity of poultry houses and yards, we were particularly impressed with those of Isaac Van Winkle, Esq., of Greenville, N. J. Mr. W. seems to have an eye to the practical utility, as well as to the beauty, of his henneries and surroundings. We present two engravings, one of which gives an interior view of the house as it is, with the exception that it is divided into sections for different classes of fowls by woven wire partitions; the other gives the south elevation of the house, and shows the interior of one of the yards. The partitions in the house correspond with the size of the yards. The building is nearly seventy-five feet long,

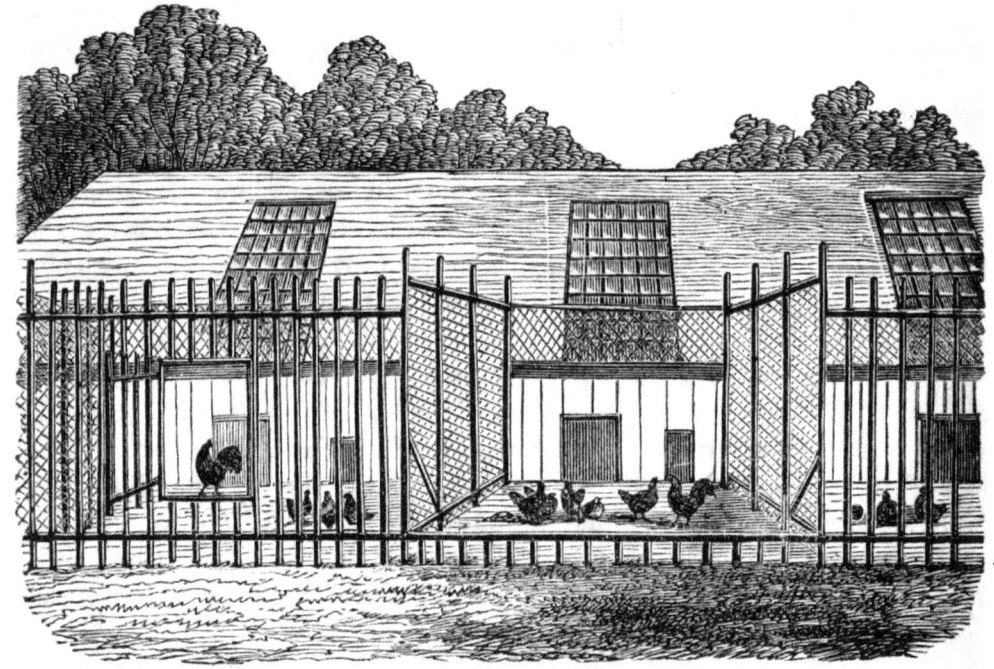

POULTRY HOUSE AND YARDS — SOUTH ELEVATION.

thirteen feet high and twelve feet wide. It is built of wood, roof shingled. To the highest pitch of the roof it is thirteen feet. The elevation or hight from the ground or foundation in front is four feet, which cuts a twelve-foot board into three pieces; the length or pitch of the roof, in front, is twelve feet—just the length of one board, saving a few inches of a ragged end; the pitch of the rear roof is six feet, and the hight of the building from the ground to the base of the roof is just six feet, which cuts a twelve-foot board into two pieces. The ground plan and frame work are planned on the same principles of economy of timber. By this plan no timber is wasted, as it all cuts out clean; there is also a great saving of labor. The foundation of the building rests on cedar posts set four feet into the ground, to prevent action of the frost in the winter and spring. These are regarded very much better than brick or stone piers. This house contains eight pens, each of which will accommodate from twenty-five to thirty fowls; each pen is nine feet long and eight feet wide. All the pens are divided off by wire partitions of one inch mesh. Each pen has a glass window on the southern front of the house, extending from the gutter to within one foot of the apex of the roof, fixed in permanently with French glass lapping over each other, after the fashion of hot-bed sashes; they are about eleven by three feet. Each pen is entered by a wire door six feet high; from the hallway, which is three feet wide; and these doors are carefully fastened with brass padlocks.

INTERIOR OF POULTRY HOUSE.

The house is put together with match boards, and the grooves of the boards are filled in with white lead and then driven together, so as to make the joints impervious to cold or wet. On the rear side of the house there are four scuttles or ventilators, two by two feet, placed equidistant from each other, and to these are attached iron rods which fit into a slide with a screw, so that they can be raised to any hight. These are raised, according to the weather, every morning, to let off the foul air. Each pen has a ventilator besides the trap-door at the bottom, same size, which communicates with the pens and runs. These lower ventilators are only used in very hot weather, to allow a free circulation through the building, and in summer each pen is shaded from the extreme rays of the sun by thick shades fastened upon the inside, so that the inside of the house is cooler than the outside.

The dropping boards extend the whole width of the pen, and are about two feet wide and sixteen inches from the floor; the roosts are about seven inches above and over this board. They are three inches wide and crescent-shaped on top, so that the fowls can rest a considerable part of their bodies on the perches. Under these dropping boards are the nest-boxes, where the fowls lay, and are shaded and secluded. The feeding and drinking troughs are made of galvanized iron, and hung with hooks on eyes, so that they can be easily removed when they require cleaning. One can stand at one end of this long house and see all the chickens on their roosts. By seeing each other in this way the fowls are made companionable, and are saved many a ferocious fight; at the same time each kind is kept separated from the other. Each pen has a run thirty-three by twelve and fifteen feet; these runs are separated by wire fences twelve feet high, with meshes of two inches. Outside of these small runs is a large run of half an acre, and on the rear are other runs of about an acre, all of grass, so that four or five kinds can be out at large at a time in these large runs, and into which they are all let out by turns.

The house is surrounded with a drain which carries off all the water and moisture, and prevents dampness. Inside, the house is cemented all through; and these cemented floors are covered with gravel about two inches deep. The house is heated in the cold weather just enough to keep water from freezing, as Mr. VAN WINKLE is opposed to much artificial heat, and to forcing fowls to lay. At the north end is a small house or shed to protect the hens from the north winds, and the entrance is by the south, through the shed which is used to keep his feed close at hand.

The plan of this hennery is remarkable for its simplicity and hygienic arrangement. The cost of the labor and material was under five hundred dollars. The house is cleaned out every day. We were there in the hottest of last summer weather, and it smelled just as sweet as outside; we could not discover the slightest taint to the air inside. Mr. VAN WINKLE has other houses. One about fifty feet long, in which he has, on the second floor, a sitting department. This house has five pens, with an office for his poulterer. He planned all his own houses, and seems to have a quick eye to any improvement. He has succeeded most admirably in all his aims, if we judge by results.

PLAN OF POULTRY HOUSE FOR ONE HUNDRED FOWLS.

This plan requires the ceiling and sides to be lathed and plastered. The partitions are made of smooth lath or boards, and set up endways and fastened securely at both ends with a space between them of from one and a half to two inches. The nests are twelve inches wide, fifteen inches high, and fifteen inches long, and so constructed that they may be slid out at pleasure from the laying-room into the sitting-room, reserving room for a door in either case to keep the fowls separate. The doors and windows are placed so that a good draft is secured in warm weather, and plenty of light in cold weather. The perches are made portable, so that they can be moved or taken out at pleasure, to make it convenient to clean out the hennery. The length of the building is sixteen by thirty feet, which is divided into six rooms or compartments, two are laying and roosting rooms, one sitting room, and three for runways or rooms for roamage.

PLAN OF A SMALL DOUBLE HENNERY.

Those desiring to keep two distinct breeds of fowls on a village lot, and having but little room to do so, we think a small double hennery can be made to answer all purposes, in a yard one hundred and forty by thirty-five feet—one of which we have seen. It can be made very cheaply, takes up but little room, and is considered a model hennery. This lot is surrounded in the rear and one side with an ordinary tight board fence; the coops are at C, and runways R, as shown in the plan. The runways are five feet wide—that to the rear of the lot being twenty-five feet long, the coops being each five feet square; the front of each runway is lathed up like any ordinary hennery. The coops are made tight, in which are situated a row of nests at N; the roosts are at P; windows are placed at the ends, which admit the light; S, denotes the slots in the coops for the fowls to pass in and out of the runways. The runways on the side of the lot may be made the full length of the same, if desired, but twenty-five feet is sufficient runway room for seven fowls. The door to each coop is situated in the corner, D. This arrangement we think very economical, and answers every purpose for keeping two distinct breeds of fowls, in a small space. If deemed advisable, the fowls could be let out on the large plat of ground on alternate days, to allow them to get grass, and pick up such refuse as comes from the kitchen and table. It is a good plan to sift coal ashes in the hen-yards for them to wallow in; also to spade up a portion of it, so that they can, in sunning themselves, wallow in the fresh dug earth, which has a tendency to keep them clear of vermin.

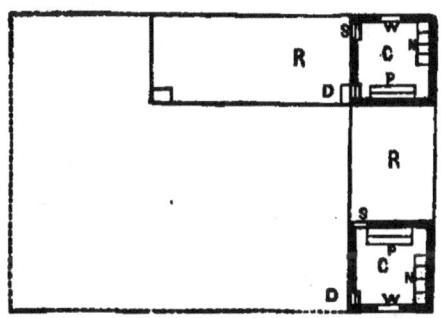

PLAN OF A SMALL DOUBLE HENNERY.

RHODE ISLAND POULTRY HOUSE.

The following plan of a poultry house is taken from the Albany *Cultivator*, and differs very considerably from those already given. The writer who furnishes the plan remarks:—"Some farmers are of an opinion that a few boards tacked together, or set against the side of a wall, answer very well for the purpose of a hen roost; but I have come to the conclusion that to render our fowls profitable, as much care must be taken of them as of our horses and cattle. This house may be built of pine boards, or it may be clapboarded and plastered with lime; in either case it should have a good

Fig. 1.

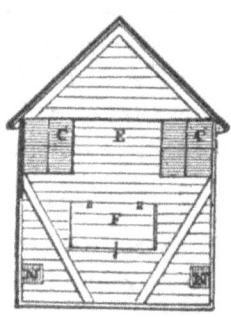

Fig. 2.

EXPLANATION.

plank floor. It is twelve feet long, eight feet wide, and seven feet high, from the bottom of the sill to the top of the plate."

Fig. 1. View of the east end; A, a door, two feet wide and five feet high; E, a small window for ventilation.

Fig. 2. View of the west end; N N, two holes one foot square for the

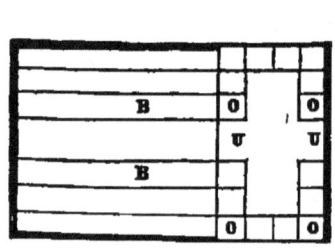

Fig. 3.

Fig. 4.

entrance of the fowls; F, a door to throw out the manure; it turns up and hooks at E; C C, windows with small wire grates.

Fig. 3. Interior view; U, a door; O O O O, boxes for nests, twelve inches square, to be placed in three tiers, one above the other; U, an inside door of the same dimensions as the outer one; B B, are poles, or roosts;

 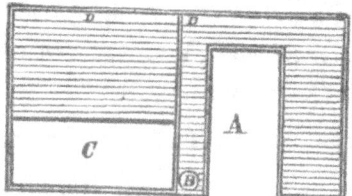 

FRONT VIEW OF A VIRGINIA POULTRY HOUSE.

these may be either of sassafras or wild cherry tree. They are fitted to swing up and hook at the upper floor.

Fig. 4. Side view; M M, nests or boxes for brood hens; these should have a long door to swing down and hook at the bottom.

VIRGINIA POULTRY HOUSE.

A writer in the fifth volume of the *Cultivator* says, "I have used the poultry house, of which drawings on preceding and this page, are represensentations, for about eight years, and can testify that it is preferable to any known in this section of country, and many of my neighbors have

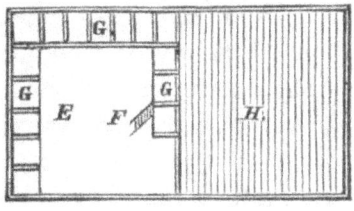
FRONT VIEW.

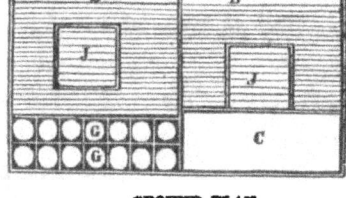
GROUND PLAN.

thrown aside their old houses and built after my plan. The roosts for the fowls should be often renewed, and always of sassafras, as the smell of that wood is deleterious to the vermin on poultry. The floor in the sitting room should always be kept perfectly clean, and continually covered with

CHEAP POULTRY HOUSE.

ashes and lime, and the litter from under the roosts taken away weekly. A, the door; B, the entrance for the fowls; C C C, the openings underneath the mitred floor, where the fowls roost; D D D, six inch openings to admit air; F, the ground floor, made of earth, elevated above the surface one foot, with boxes for the poultry to lay and sit in; F, a ladder for poultry to go to

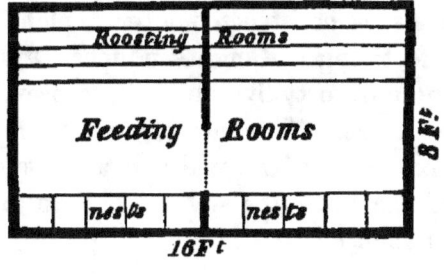
GROUND PLAN.

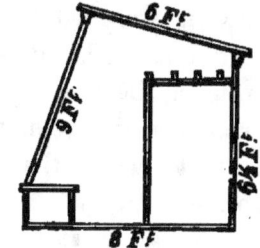
TRANSVERSE SECTION.

their roosting room; G G G G, boxes for nests; H, lattice floor for the litter from the poultry to roost in; I, a round hole, one foot in diameter, for fowls to roost; J J J, lattice windows of blinds three feet wide, and three feet six inches deep."

CHEAP POULTRY HOUSE.

In third volume of the *Country Gentleman* we find the plan, on page 115 with the annexed elevation, of a cheap poultry house, furnished by a correspondent. He says:—"I have thought it would not be out of place to send you a drawing and plan of one we consider the best, as it can be made to accommodate from one dozen to five hundred fowls. The plan I send is sixteen feet long by eight feet wide at the bottom, and costs, using one-inch matched boards, about one dollar per foot. The present one will cost from sixteen to twenty dollars, including sash, doors, and other fixtures. The engraving exhibits the plan so clearly that any explanation is altogether unnecessary."

VAN NUXEN'S POULTRY HOUSE.

"Having made some experiments in the raising of chickens, a business that forms a part of every farmer's occupation, I send you a description of my present plan of operation, which appears to answer admirably. Under an outhouse, sixteen by eighteen feet, raised three feet above the ground, I have made a cellar three feet below the ground, making the hight six feet altogether. Eight feet in width of this cellar is partitioned off for turnips, the remaining ten by sixteen feet being sufficiently large to accommodate one hundred chickens, or more. This cellar is inclosed with boards at present, but it is intended to substitute brick walls in a year or two. The roost is made sloping from the roof to within eighteen inches from the ground or floor, twelve feet long by six feet wide. The roost is formed in this way: Two pieces of two-inch plank, six inches wide, and twelve feet long, are fastened parallel, six feet apart, by a spike or pin, to the joist above, the lower end resting on a post eighteen inches above the ground. Notches are made along the upper edge of the plank, one foot apart, to receive sticks or poles from the woods, the bark being left on. When it is desirable to clean out the roosts, the poles, being loose, are removed; the supports, working on a pivot, are raised and fastened up, when all is clear for the cleaning out. I next provide the fowls with corn, oats and buckwheat in three separate apartments, holding about half a bushel each, which are kept always supplied. A row of nests is constructed after a plan of my own, and does well. It is a box, ten feet long and eighteen inches wide; the bottom level, the top sloping at an angle of forty-five degrees, to prevent the fowls roosting on it; the top opens on hinges. The nests, eight in number, are one foot square; the remaining six inches of the width is a passage way next to the wall, open at each end of the box; the advantage is to give the hens the apparent secrecy they are so fond of."

OCTAGON POULTRY HOUSE.

Those desirous of keeping from twenty to thirty fowls will find the octagon style of a house just the thing for them. It is more ornamental than the oblong house and economizes room, where that essential is required.

The object of placing it on piles is to prevent the encroachment of rats and other animals that prove so destructive to eggs and fowls, when not properly protected. The structure is not a costly one; any person used to handling tools can construct it, at a merely nominal expense, adding ornamentation to the structure, as he desires. This building is ten feet in diameter and six and a half feet high. The sills are four-by-four and the plates three-by-four joists, halved and nailed at the joints. It is sided with inch-and-a-quarter spruce plank, tongued and grooved. No upright timbers are used. The floor and roofing are of the same kind of plank. To guard against leakage by shrinkage, the joints may be battened with laths or other strips of thin board. An eight-square frame supports the top of the rafters, leaving an opening of ten inches in diameter, on which is placed an octagon chimney for a ventilator, which makes a very pretty finish. The piers should be either cedar, chestnut, or locust, two feet high, and set on flat stones.

OCTAGON POULTRY HOUSE.

The letter D, designates the door; W, W, windows; L, latticed window to admit air, with a shutter to exclude it when necessary; E, entrance for the fowls to alight on when going in; R, R, are roosts placed spirally, one end attached to a post near the center of the room, and the other end to the wall; the first or lowermost one two feet from the floor, and the others eighteen inches apart, and rising gradually to the top, six feet from the floor. These roosts will accommodate forty ordinary sized fowls. F, F, is a board floor, on an angle of about forty-five degrees, to catch and carry down the droppings of the fowls. This arrangement renders it much more convenient in cleaning out the manure, which should be frequently done. The space beneath this floor is appropriated to nests, twelve in number, fifteen inches wide, eighteen inches deep, and eighteen inches high. In order to give an appearance of secretiveness, which it is

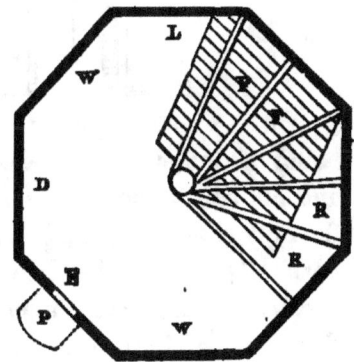

GROUND PLAN.

well known the hen is partial to, the front is latticed with strips of lath. By this arrangement a free circulation of air is admitted, which adds much to the comfort of the hens while sitting. In the foregoing bill of lumber for building purposes, spruce is given, but any other lumber convenient and as hand will of course do full as well for the structure. If the lumber used be unplaned, paint the building inside and out with either hot lime made to the consistency of whitewash, or common paint of the color which most suits the fancy. The paint or whitewash not only beautifies the building, but preserves it.

PLAN OF CHARLES MOUNT'S HEN HOUSE.

This house can be cheaply constructed, and has the advantage of being easily kept clean, as the droppings fall on the inside roof, or slide, under the roosts, and can be scraped down into the passage ways, (A) and swept out at the doors, (B,) in which are the smaller doors, (C,) hung from the bottom, and swinging outward and downward at an angle, to allow the fowls to enter, at the same time keeping out rats and other vermin—the outer end

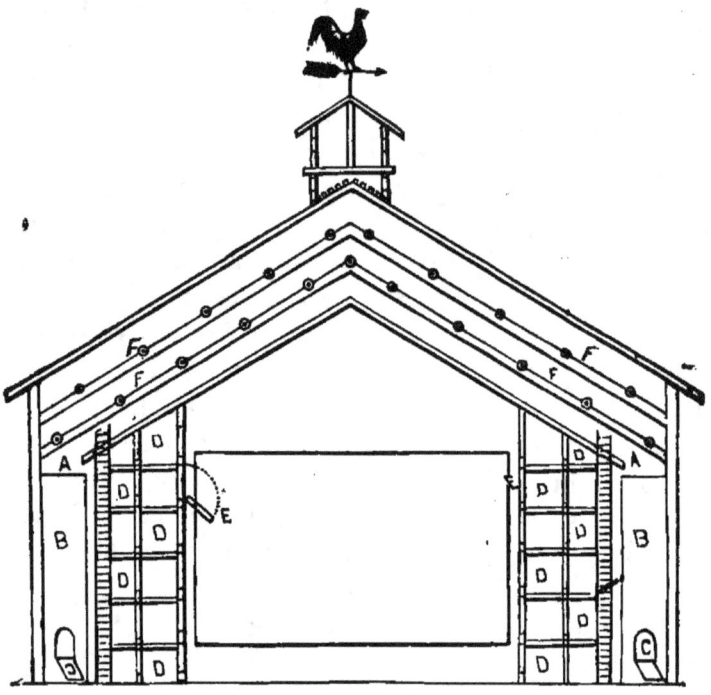

PLAN OF CHARLES MOUNT'S HEN HOUSE.

being about six inches from the ground. This door can be closed at night and in cold weather. The nests are ranged in tiers on each side of the feeding room, the hens having access to the nests (D) by ladders running vertically across the face of the platforms, which also give access to the roosts. This arrangement of the nests (D) gives the fowls privacy and darkness, and allows them to follow their inclination to steal away and hide their nests. The nests are easily got at to remove the eggs or clean them out, by opening the boards, (E,) which run the whole length of the tier of nests, and are hinged at the bottom side and held, when closed, by a button at the top. There is a door, (E,) at each side, at the end of the tier of nests, opening from the feeding room to the passage on each side, which also has an opening in the bottom for the passage of the fowls, fitted with a small door to shut them out when necessary. The roosts (F, F,) are round poles, which rest in notches cut in pieces which are fastened to each end of the building, which allows of their being taken out to clean. The gable end should face to the sun, and have double sash covering the whole size of the feeding room down to within one foot six inches of the ground, to let in the light and heat of the sun in winter. The roof at the peak is left open for ventilation, and surmounted by a double row of pigeon boxes, the under

POULTRY HOUSE — ELEVATION.

side of which have boards hung to close in extreme cold weather. The whole is surmounted by a vane to give it finish. The house is eighteen by thirteen feet, and eight feet post; is clapboarded outside and ceiled inside with worked boards, and filled in with tan bark. It can be floored with plank or cemented.

PLAN OF POULTRY HOUSE THAT WILL ACCOMMODATE ONE HUNDRED FOWLS.

A yard fifty by one hundred feet is sufficiently large to answer the purpose desired by a medium breeder, and upon which one hundred fowls

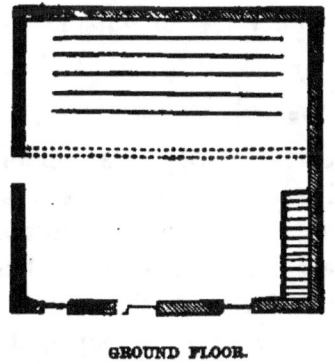

GROUND FLOOR.

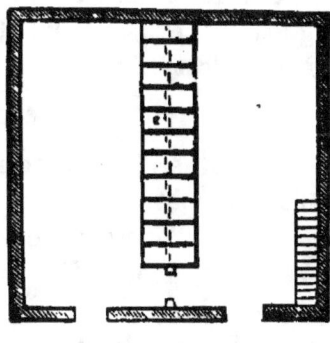

SECOND FLOOR.

can be conveniently kept. But the more room allowed them the better it will prove for the health of the brood. This lot should be allowed the fowls outside of the dimensions of the hennery. We have seen a flock of one hundred fowls well kept upon the space mentioned. A poultry house, containing two floors, constructed on the following plan, which we take from the *Scottish Farmer*, will accommodate, very comfortably, one hundred fowls during winter and summer, provided they are allowed the liberty of roaming in a small yard during pleasant weather. The cost of the building, of course, will vary according to style of construction and price of materials. The house is considered, in England, a desirable one, and answers the purpose so well that it is being extensively used by poultry fanciers of limited means. The plan presents some features of novelty as well as of utility. The posts of the frame, if built of wood, may be not over nine feet high, by resting the sills on concrete walls of three feet, where it is convenient to build on a slight inclination. Seven and a-half feet in hight will do for feeding room and the manure pit, which may be formed by running a wall three feet high, as shown by the dotted line. The manure may be thrown in through the door, which opens near. Three windows on the south side will give light and warmth. The second floor may be lathed up the roof, to give sufficient hight in the center, which will be four and a-half feet under the eaves of the roof. The nests are set in the partition, one foot from the floor, one foot high, and one and a-half feet long, open at both ends with a slide door, which is reversed when a hen is sitting, so that she is placed in the opposite or sitting room, and thus the others never disturb her. A door

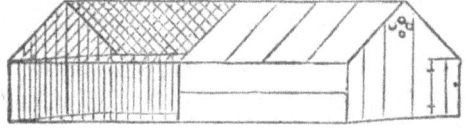

CHEAP POULTRY HOUSE.

to communicate between the rooms and windows in the end and south side will give sufficient light and warmth. The whole may be lathed and plastered, or ceiled up, either of which will make it a warm and durable building. The lower floor is much the best for roosting and feeding, while the hens can quite readily ascend to lay and sit on the upper floor with equally good success. "This plan will give eighty feet of outside wall and eight hundred feet of floor."

CHEAP POULTRY HOUSE.

Here is another house, designed to meet the wants of a person who cannot afford to put up a poultry house, and who has but a small yard for fowls to run in. The figure illustrates the design of the house. On the right is the house, with door. The house is four feet long, three and a half feet wide, twenty inches high at the sides, and thirty inches at the peak. Inside are a roost and a couple of nests. In the rear of the yard a coop is attached to the house, as shown in the drawing, in lattice work. It is five feet long, and the same width, hight, and shape as the house. The house opens into the yard by a hole a few inches from the ground; it is ventilated by a few auger holes bored in each end in the peak. A pane or two of glass may be put in, if desirable. This coop can be moved daily, so that the fowls will be on fresh ground. It will accommodate a cock and six hens. For breeding purposes, where it is desirous to coop up a particular trio or more, it is invaluable.

A HENNERY THAT WILL ACCOMMODATE FROM TWO HUNDRED TO TWO HUNDRED AND FIFTY FOWLS.

Mr. G. O. Brown of Maryland, gives the following as a plan of a cheap and convenient hennery, for those wishing to keep a large number of fowls. He says:—"The drawing of the building shows the north and west sides. The building is sixteen by twenty feet, sixteen feet high to roof peak. Fig. 2 represents the inside of the building as follows:—C, roosting and general room; B, egg room, feed room, etc.; A, A, A, are nests. In the recess there are three rows of nests, one above the other; 5, door opening from outside building; 6, door opening from feed room to recess, nest boxes and roosting room.. Fig. 3, nest boxes, thirteen by twenty inches. These boxes are all movable, so that I arrange them to suit circumstances. By raising a board, hinged, one can readily examine the nests from the feed or egg room. Should a hen wish to sit, take out one of the nest boxes, turn it end for end, thereby placing the end that is closed up in the roosting room, which prevents the other hens from bothering or annoying her. I have it so arranged that the sitting hen can go out in a little yard, scratch and dust without any inconvenience or annoyance from the others. The egg or feed room has shelves in it, and a loft, (which is reached by a ladder made fast up the side,) where the feed is kept. Fig. 4 represents the roosts, two feet apart, of sassafras. Fig. 5 represents a flooring of boards, with the same slant as the roosts, but placed two feet away from the roost. The droppings falling on these boards, roll down into a trough at the lower end, as shown. In the east side of the house I have one large sliding window, and in the south side two, with wire fenders or screens, over all three. A building of this size and kind can accommodate two hundred to two hundred and fifty chickens with ample room.

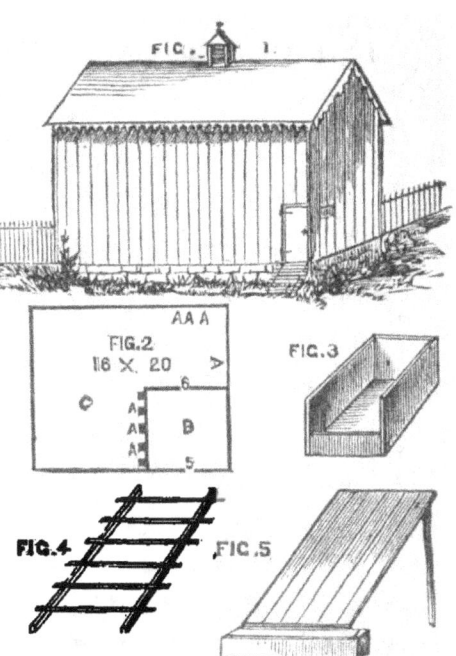

A HENNERY THAT WILL ACCOMMODATE 250 FOWLS.

PLAN OF POULTRY HOUSE THAT WILL ACCOMMODATE THREE DISTINCT BREEDS.

The building is enclosed with worked spruce or pine boards, put on vertically, and the hight so arranged that each board will cut to avoid waste. Elevation — Length, twenty-four feet; width, eleven feet; hight in front, nine and a half feet; hight in rear, six and a half feet. All the

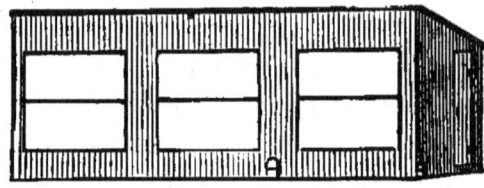

POULTRY HOUSE — ELEVATION.

pieces are cut off of the full lengths in front, making just half a rear length. The rafters, of thirteen feet joist, with either battened or shingle roof as preferred.

PLAN AND YARD.

The building is supposed to face the south. The entrance door, E, opening into the passage, P, three and a half feet wide, which runs the length of the

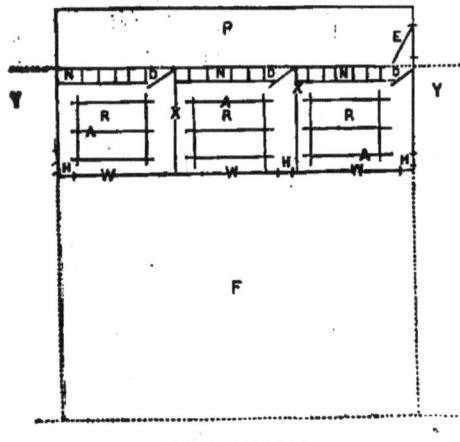

PLAN AND YARD.

building; smaller doors, D, each two feet wide, opening into the roosting rooms, R. The nests are raised about a foot from the floor, and also open into the rooms R, with a hinged board in the passage so that the eggs can be removed without entering the roosting rooms. The perches, A, are movable, perfectly level, and raised two feet from the floor. The partition walls are tight, two boards high, above which is lath; the passage wall above the nests, and the doors, D, D, D, being of lath also. The roosting rooms are seven and a half by eight feet, large enough for twenty-five fowls each. Windows are six feet square, raised one foot from the floor. We prefer the glass to be six by eight or seven by nine inches — as these small sizes need no protection strips to prevent the fowls from breaking them. The holes, H, for egress and ingress of the fowls, are closed by a *drop* door worked by a cord and pulley from the passage way. Another door can be placed in the other end of the passage way if desirable.

This arrangement of the yards, Y, of course would not suit every one; some would prefer smaller yards, making each yard the width of the room and adding to its length. We can only say "cut your garment according to your cloth" — cut your yard according to your ground. The house above is designed for only three varieties; but by simply adding to the length, any number of breeds may be accommodated. The simplest and most economical foundation is to set locust or oak posts about four feet deep, every eight feet, and spike the sills on them. There is then no heaving from frost; and all the *underpinning* necessary is a board nailed to the sill and extending into the ground a couple of inches. A sitting room can be added by making the building four feet longer. The room should be in the end next the door, so as to be always within notice.

MR. HAWLEY'S POULTRY HOUSE.

Mr. HAWLEY gave in the *Rural New-Yorker*, a few years since, a plan and description of a poultry house which he said proved a success with him during severe cold weather—the thermometer indicating only three degrees below freezing, when it was fifteen degrees below zero outside. The house is twenty feet long, eight feet wide on the bottom, six feet high in the rear, six and one-quarter feet in roof. It is built of matched and dressed lumber

OCTAGONAL POULTRY HOUSE.

or the outside, battened with strips and well painted. The frame is three by four inch joist—lathed and filled in with sawdust on all sides and roof,

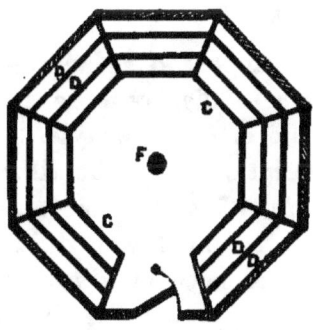

GROUND PLAN.

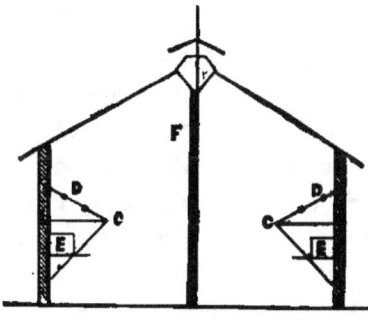

END SECTION.

then plastered. Gravel bottom. There are three windows, twelve lights, nine by thirteen, both sashes movable, and a light frame, one-half the size of the windows, prevents the fowls from escaping when the sashes are raised or dropped. The building is divided into three comfortable coops. There is ample room for two lengths of roosts, under which there is a platform to catch the droppings, thereby insuring cleanliness, so essential to the health of the occupants. The nests are "secret," built on the ground. A window in the door regulates the temperature of the house.

OCTAGONAL POULTRY HOUSE.

The *Canada Farmer* gives the following plan for an octagonal poultry house, and says " that form is chosen as offering a greater internal space for the same extent of wall than the square form. The door occupies one of the sides, the windows two of the others. The roof is supported by a central pillar, F, and, if desired, may have a lantern light at the top, with louvre boards, or other openings, for ventilation. The center pillar is by far the best plan of supporting the roof, for if horizontal tie-beams are used, the fowls will unquestionably perch on them. Around seven sides of the interior runs a broad, stout shelf C, C, over which the two lines of perches D, D, are supported on inclined rests. Underneath C, C, is a narrower shelf for the nest boxes E, E. If desired, movable baskets or boxes can be placed on this shelf.

"The advantages of this arrangement are obvious. The fowls, following the natural instinct which leads them to select the highest perches, roost over the shelf, and the nest boxes are undefiled. The manure on the shelf is in a position in which it can be easily scraped away with a flat hoe or scraper, and the shelf sanded daily. The floor is kept free from filth, and the house consequently always preserved clean and wholesome. The space under the nest-boxes will serve for the cooping of the hens with chickens, if no better situation offers. If extreme cheapness of construction be an object, the house may be built by driving light poles into the ground at equal distances, and closing in the spaces between them with weather boarding. The form admits of easy ornamentation, and may be adapted to harmonize with almost any style of buildings."

A FANCY POULTRY HOUSE.

In closing our remarks on poultry houses, we cannot do so without presenting to those who can afford it, and who wish to display more taste in this branch of economy, an illustration of a Gothic or Chinese style of poultry house, which we take, together with the description, from the *American Poulterer's Companion*. It is a very neat and pretty looking structure, and is designed for a poultry house and yard for breeding fowls, ducks and pigeons. It is intended to stand in the center of a piece of grass land or park, and if on a slight knoll or mound so much the better. If the soil is inclined to clay, it should be excavated all around the building at least two feet deep, and first a layer of stones about one and a half feet deep, then covered with coarse gravel and sand. This is desirable — for we consider it almost essential to success — stagnant moisture or wet in the soil being more conducive to disease than any other circumstance. A southern aspect is the best, and if sheltered from the north and northwest, by plantations of evergreens, it will not only be a protection from the cold winds of winter, but a shelter from the rays of the sun in summer. The houses and yards must be constructed to suit the views and purposes of the proprietor. The yards should be fenced with pickets at least six and a half feet high—wire would be more ornamental but rather expensive. Not less than one-fourth of an acre should be allowed for fifty fowls. The walls of the poultry house should be of brick, nine inches thick, and hollow; they should be at least twelve feet high, so that the roof can project some four feet, forming a shed for protecting the fowls from the storm. The front of the shed may be formed

FANCY POULTRY HOUSE.

of lath or any other kind of wood, in a rustic manner, forming a trellis on which vines might be trained, which would add much to its appearance; or it may be inclosed with glass, and grapes grown on the rafters; or nests may be placed in these sheds for sitting hens.

We may observe here, that whichever plan is adopted, the cheapest and warmest materials of which to construct the house are a wood frame and a weather-boarding, either of clapboards, or ceiled up and down with narrow battens. It should be ceiled within with hemlock boards, tongued and grooved, and laid crosswise, and filled in between the timbers with spent tan, or any other dry substance, well rammed or packed in. Or the spaces between the posts may be filled in with brick and a thin coat of plaster. In either case, whether of brick or wood, it should be whitewashed with lime. The roof should also be ceiled with boards and filled in with tan, which would render it cooler in summer and warmer in winter. The interior may be finished to accommodate the kind of stock intended to be kept. If for the large Asiatic fowls, the perches should be low, or the floor of their roosting room may be covered with straw; in which case it should be cleansed, or the straw changed daily. The cupola is intended for a pigeon house. The holes by which they enter should not be too large or numerous, and should have a shelf at the entrance. The upper tier should have a roof or weather-boarding over them to keep out the wet. An objection to a wooden pigeon house is, that they are too cold in winter and too hot in summer; but this may be in a great measure prevented by making the wood double, with a space of two or three inches between, which will form a non-conductor of heat.

POULTRY APPLIANCES.

CHICKEN-COOPS OR PENS, FEEDING-HOPPERS AND TROUGHS, WATER-FOUNTAINS, ETC.

As the rearing, management and care of all kinds of poultry is much facilitated, as well as rendered more certainly remunerative by the aid of suitable appliances, we give herewith engravings and descriptions of those we have been enabled to cull from the sources at command, and which are deemed every way suited to the ends desired.

CHICKEN HOUSE — EXTERIOR VIEW.

The march of improvement in the building of chicken houses seems to be as manifest as in most other things, and anything new in this line is sought after with interest by the amateur or breeder of fancy fowls. We give an illustration (two views) of one of these houses, which struck us as being the *ne plus ultra* of chicken coops. It is sketched from coops we saw on the grounds of Isaac Van Winkle, Esq., of New-Jersey. One illustration shows the house with the end open, giving an interior view, while the other shows the exterior. These houses are movable; made of matched boards nailed to posts, two by two inches, on each end, and side or section, and hooks and staples placed at the top and bottom of the posts, on each inside, so that instead of being nailed together as a whole, it is hooked up in sections, as shown in the engraving, with front section down. By this means the coop can be taken down and moved to any place desired. At the rear of the runway is placed a tight coop, as shown, into which the hen and chickens can retire to roost; the slide being closed, makes it perfectly rat or vermin proof. In the end of the tight coop are three or four one-inch holes

CHICKEN HOUSE — INTERIOR VIEW.

made for ventilation. The top of the runway is covered with a movable glass sash — hot-bed fashion — under which chickens can be reared in the coldest weather. For ventilation, the sash can be slid off, as seen by reference to the cut, or, if desired, the sash being placed in a groove, can be removed entirely from the top of the coop. It strikes us that this house, when it becomes more known, will somewhat revolutionize the rearing of early spring chicks. It is so constructed that any person at all conversant with the use of tools can put one up in short order and with comparatively little expense.

It is frequently recommended to breeders to build their chicken-coops with floors in them. We cannot see any particular benefit derived from having coops with wooden floors; on the contrary, we are of the opinion it inclines the chicks to weakness. Our mode is to let the chickens have free access to the ground, or, what is better, let the coop be placed over a flooring of ashes, made about two inches thick, so that the mother-hen can dust herself at pleasure.

THE RAT-PROOF COOP

is our *beau-ideal* of what a chicken-coop should be. It can be moved at will, and at evening, or in stormy weather, the hen and chickens can be driven in and the coop closed up, making it both rat and water-proof. Then again, there is

THE TENT COOP,

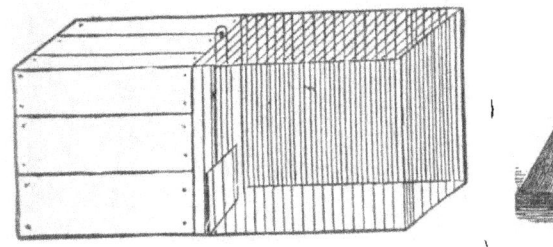

THE RAT-PROOF COOP.

THE TENT COOP.

which answers a good purpose for either young turkeys or chickens; is easily constructed, and, having no floor, can be moved to any light or sandy soil, which will answer in lieu of ashes for the chicks and hen to dust themselves in, which keeps off vermin.

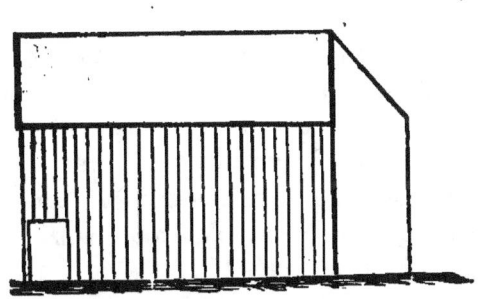

THE PENT OR LEAN-TO COOP.

THE PEN COOP.

There is another form of coop which we have, but it has no advantages to our mind over either of the preceding ones, unless it be that in this shape it affords more room. It is made of clapboards and lathed or lattice-worked across the whole front of the coop and about half-way up as shown in the engraving. The back, sides and top are made of clapboards, but any kind of boards might be used. We used the clapboards for the sake of making the coop light, so that it could be handled easily. The size is four and a half feet long — two and a half feet wide — front three feet high and with a pitch of half a foot to the rear; the front of the coop being clapboarded one foot down, leaving two feet for the length of the laths in front. The clapboard in front has the effect to prevent the storm from beating in upon the hen and chicks which are confined within the coops.

PEN CHICKEN COOP.

The illustration, as shown on page 129, gives a good idea of a pen coop, which may be made large enough to contain a cock and four hens for breeding purposes, where they can enjoy the sun and fresh air, yet be protected from stormy weather. The dimensions are as follows:—Pen four feet high in front and three feet in the rear, six feet long and four wide. The yard, ten feet long and six wide, to be enclosed with lath four feet high. If desired, the top may be covered also. The pen may be made with common

THE BARREL COOP.

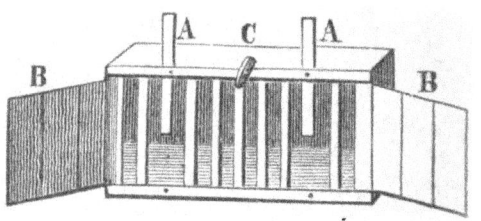

THE CLOSE COOP.

boards, and battened up as shown in the cut. The holes in the ends are made to admit a free circulation of air. This house may, of course, be reduced to a size sufficient to keep a hen and chickens. If used for that purpose the coop may be say twenty inches high in front, fifteen in rear, and twenty inches square on the bottom, making it light and movable.

THE BARREL COOP.

This is an appliance to keep chicks, that any one can make, with very little trouble, as will be seen by our illustration. All that is necessary is to place a common flour or other old barrel on its side, take out one of the heads, place some straw in the back end of the barrel; put the hen and chicks in; have some lath or strips of board at hand, with one end sharpened and drive them into the ground in front of the coop and your work is accomplished.

THE CLOSE CHICKEN COOP.

This coop is very handy, and may be made of inch boards (see illustration,) long enough to admit of any number of fowls. A, A, are slats raised for admitting the fowls; B, B, doors to open and shut at night, to prevent the intrusion of any kind of vermin; C, button for fastening the doors. Any common dry goods or other large box will answer the purpose desired. Cleats can be nailed on, as shown in the engraving, which makes it a light, warm and airy coop in summer. Holes should be bored in one end and in the top for ventilators.

FEEDING HOPPERS AND TROUGHS.

In giving descriptions of the different varieties of Feeding Hoppers or Troughs being used by poulterers in their endeavor to facilitate the workings of domestic economy in the poultry yard, we cannot do better than to commence by presenting for inspection and adoption, if desired, a plan of the Scotch Feeding Hopper, taken from *Loudon's Encyclopedia of Agriculture,* the description of which is given below.

A SCOTCH FEEDING HOPPER.   A PERFECT FEEDING HOPPER.

THE SCOTCH FEEDING HOPPER

can be made to contain any quantity of corn required, and none wasted. When once filled it requires no more trouble, as the grain falls into the receiver below as the fowls pick it away; and the covers on that, which are opened by the perches, and the cover on the top, protect the grain from rain, so that the fowls always get it quite dry; and as nothing less than the weight of a fowl on the perch can lift the cover on the lower receiver, rats and mice are excluded. In this connection we give an engraving of what is called

A PERFECT FEEDING HOPPER,

in *Bement's Poulterer's Companion*, which we think superior to that of the Scotch plan, and which, from the description here given, can be easily constructed by any person. "A is an end view, eight inches wide, two feet six inches high, and three feet long; B, the roof projecting over the perch on which the fowls stand while feeding; C, the lid of the receiving manger raised, exhibiting the grain; E, E, cords attached to the perch and lid of the manger or feeding trough; I, end bar of the perch, with a weight attached to the end to balance the lid, otherwise it would not close when the fowls leave the perch; H, pulley; G, fulcrum. The hinges on the top show that it is to be raised when the hopper is to be replenished. When a fowl desires food, it hops upon the bars of the perch, the weight of which raises the lid of the feed box, exposing the grain to view, and after satisfying its hunger jumps off, and the lid closes." Of course the dimensions of either of these feeding hoppers may be increased to any size desired.

STANDARD SELF-FEEDING HOPPER.

This feeding hopper is two feet square, the posts eighteen inches high, three inches square; the upper section of the box is six inches deep, the ends are mortised into or nailed to the posts. From the bottom of this square the tapering part of the grain box reaches to within one inch of the floor, which should be raised on feet about six inches from the ground; the grain box tapers to one foot square, and to bring the grain within reach of the fowls, a cone, as shown at A, is placed in the center of the floor, and should be so much smaller than the funnel part of the hopper as to leave at least one inch space all

STANDARD SELF-FEEDING HOPPER.

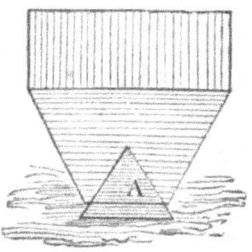

FUNNEL AND CONE.

around the cone, which forces the grain to the edge, where, as the fowls pick the grain away, more will fall and keep a constant supply within reach of the fowls, as long as any is left in the hopper. The slats on the sides are intended to prevent the fowls from getting into the trough or crowding one another. This hopper will hold about two bushels of grain, and if the roof projected one foot all round it, it would protect it completely from rain. It occupies but little space, and from twelve to sixteen fowls can feed at the same time.

A STOOL FEEDING HOPPER.

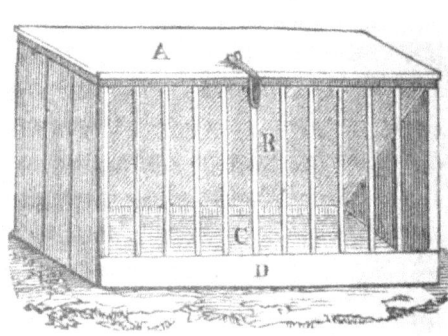

A CHEAP FEEDING HOPPER.

A STOOL FEEDING-HOPPER,

which is proof against rats, can be made as follows:—Make a platform two or three feet square, as the case may be; then make a square box, three inches high and sixteen inches square; nail it in the center of the platform; saw strips one and one-fourth inches square and eighteen inches high for the posts; nail strips\of boards, two inches wide, to the posts at top, to secure and steady them; then take common lath, or any thin stuff, one and one-half or two inches wide, and nail them to the top and bottom, up and down, leaving a space of two inches between each slat, so that the fowls can get at the feed. The roof may be four-square, as shown in the engraving, and detached so that it can be raised when required to be replenished with grain. Elevate the hopper on a post about three feet from the ground, as shown in the cut, which makes it rat and mice proof. The fowls will soon learn to leap upon the platform, and feed from the grain box between the slats.

A CHEAP FEEDING-HOPPER.

There is a cheap plan for a Feeding-Hopper, which can be made out of an old candle-box, for the want of a better thing. Take off the lid and one of the sides; let the ends, bottom and one side remain; cut a small strip off one end of the lid, so that it will slip in between the ends of the box, placing the lower edge one and a half inches from the side and an inch from the bottom; the other edge of the lid is to reach the top and outside corners of the ends, thus forming a deep, angular box, with long aperture at the bottom. As shown in the cut, the lid forms the slanting side B; C, forms the trough, where the corn will descend down to it when put into the angular box; then put hinges on the lid, A; the open part of the hopper has a row (D) of slats two inches apart; these slats should be brought to the edge of the box, so that the fowls can just reach the bottom of the angle; the corn falls down as fast as the fowls pick it away.

DOUBLE FEEDING HOPPER.

This hopper was highly commended by the late N. C. BEMENT. It is nine feet long and nine inches wide; end pieces fourteen inches high, and the bottom raised six inches from the ground; the ends nailed to the bottom, and a strip of board four inches wide was firmly nailed on the sides, raised three inches above the bottom board, forming a manger or trough to prevent any waste of food. Another strip of board three inches wide was nailed on the top in front to secure the ends. The hopper to contain the grain was formed of two pieces of board, nine inches wide, set between the ends forming a V, the upper edges lying against the front top strips and the bottom resting on some small blocks, from one

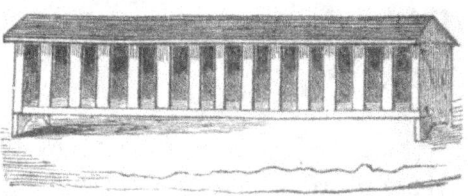

DOUBLE FEEDING HOPPER.

to two inches high, sufficient to allow the grain to fall down as the fowls pick it away. It may be made to open and shut at the bottom to suit the different sizes of grain. The top or roof may be made of the same width as the box, or it may extend over the sides sufficiently to protect the fowls from rain when feeding. Narrow strips of lath must be nailed to the top and bottom pieces, leaving space enough between them for the fowls to enter their heads when eating. It is open on both sides, and one of this size is sufficient for seventy-five fowls.

FEEDING TROUGHS.

It has been frequently suggested that in feeding fowls soft food, instead of throwing it upon the ground, thereby wasting the larger portion of it, a dish or trough of some sort should be placed in the hennery or coop to contain the food. In figure 1 we give an engraving of a trough that may be

Fig. 1.   Fig. 2

procured at little cost, which will meet the wants of most breeders. It can be made of zinc, tin, or earthenware, in an oblong form, to any desired length — width four inches, and two to four inches deep. To prevent the chicks getting into the trough and scratching the food out, a loose curved cover, made of tin, zinc, or wood, in form as seen in figure 2, will answer the purpose. The wires which support the cover should be perpendicular, ten to twelve inches high, and set two and a half inches apart. One end of the wire may be driven into the ground, if desired, for a stationary feed box, instead of having the top and bottom of the cover soldered to the wire. This trough can be made very cheaply by any tinsmith, and will economize food enough during one season to pay for more than a dozen such troughs.

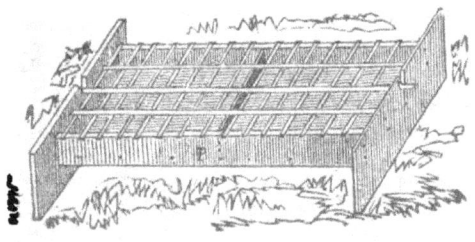

Fig. 3.   Fig. 4.

Figure 3 is a trough divided by the partition P. If desired, one part may contain water and the other food, the whole being covered with a screen made of lath nailed together in the form shown, and kept in position by the ends of the center laths fitting in and keyed to the ends of the trough. The bottom is square or of triangle form. This is for the use of grown poultry. The lattice work prevents them getting into the food with their feet. A cheap and substantial feed or water dish is shown in figure 4. The sides of an old tin pan are connected together by a net-work of wire passing through holes beneath the rim, and crossing above the center at the various angles.

DRINKING FOUNTAINS.

Figure 1 shows a barrel fountain; it has a small tube extending from the cask to a shallow dish or pan, which should be small, so that the fowls cannot get into it and soil the water. Figure 2 shows a bottle fountain, which may

Fig. 1.—Barrel Fountain.

Fig. 2.—Bottle Fountain.

be made by taking a two or three inch plank and scooping it out one and a half inches, forming a shallow trough; then make a frame similar to the figure shown, and insert the neck of the bottle, the nozzle reaching to within three-quarters of an inch of the bottom of the trough. Either of these designs will answer all purposes of a drinking fountain for the poultry yard.

THE ORDINARY POULTRY FOUNTAIN

is too well known to need description, but a rather better form than is usually made is shown in figure 3. The advantages of such a construction are two;

Fig. 3.

Fig. 4.

the top being open, and fitted with a cork, the state of the interior can be examined, and the vessel well sluiced through to remove the green slime which always collects by degrees, and is very prejudicial to health; and the trough being slightly raised from the ground, instead of upon it, the water is less easily fouled. But either form, if placed with the trough *towards* the wall, at a few inches distance from it, will keep the water clean very well. Some experienced breeders prefer shallow pans; but if these be adopted they must be either put behind rails, with a board over, or protected by a cover, in the same way as the feeding troughs already described.

WINTER WATER FOUNTAIN.

We are indebted to the *American Agriculturist* for the description and manner of keeping water from freezing in the fountain in winter, also for a duck-feeding contrivance, and for the plan of a beautiful rustic duck coop, herewith given. In describing these appliances it says:—"There is almost

FIG. 3.—FEED BOX FOR DUCKS.

always some difficulty in keeping fowls supplied with water in cold weather. We have had no trouble since adopting the following expedient. A barrel is sawed into two tubs, and an earthen jug placed in one of the tubs, the bottom of the jug and that of the tub being in contact, or nearly so, and the mouth of the jug close to the rim of the tub. The jug may be fixed in position by a few sticks, nailed across the tub inside. The tub is then stuffed full of horse litter and manure, and strips nailed across the top to keep it in. When this is done we fill the jug with water, put in a cork, and invert tub and all. (See figure 2.) Then the cork is withdrawn at the same time that a small pan is slipped under. The pan remains full during the day, and, if set in the sun, will not freeze so much as a film of ice upon the surface, even out of doors, except on the severest days. At night the pan should be withdrawn, and the water allowed to flow out.

CONTRIVANCE FOR FEEDING DUCKS.

"A simple contrivance for feeding ducks and not allowing chickens to share their food, was shown us lately at the yard of a subscriber, and we have had it engraved (figure 3.) The food was placed in a square, flat pan, in which a few bricks were laid, filling the middle of the pan, to prevent the food being shoved beyond the reach of the ducks. Then a box was turned over the pan and contents, and supported upon a brick at each corner. After a little experience the ducks learned to run their flexible necks under and fill themselves, while the disconsolate hens could get nothing. Ducks will increase rapidly in weight if they have all the soft food they can eat. The best place for them to pass the nights in winter is upon a fresh manure heap, under cover. If one wishes to feed chickens and not ducks, a convenient way is to lay a board or two, to put the feed on, upon two barrels or wooden horses.

Fig. 4.—Duck House.

RUSTIC DUCK HOUSE.

"In figure 4 we have represented a rustic, bark-roofed duck coop, which might be used either to confine an old duck and her brood at night, provided the slat-work was so close as to prevent the entrance of rats or weasels, or to confine a hen with a brood of ducklings, in which case the openings would need to be larger, and the coop would have to be shut up at night by a close front. There is more danger to young ducks from rats than from any other cause."

WIRE-COVERED RUN.

We find in the *Practical Poultry Keeper* what is termed therein a wire-covered run for chickens, and is deemed by the author a sensible way to keep the young chicks from being destroyed by cats or rats. The plan strikes us as being one worthy of consideration; we therefore give it space

in our pages, together with the remarks accompanying the engraving:—
"Cats sometimes make sad inroads on the broods. If this nuisance be great it is well to confine the coveted prey while young within a wire-covered run. And the best way of forming such a run is to stretch some inch-mesh wire-netting, two feet wide, upon a light wooden frame, so as to form two wire hurdles, two feet wide and about six feet long, with one three feet long. These are easily lashed together with string to form a run six feet by three, and may be covered by a similar hurdle of two-inch mesh three

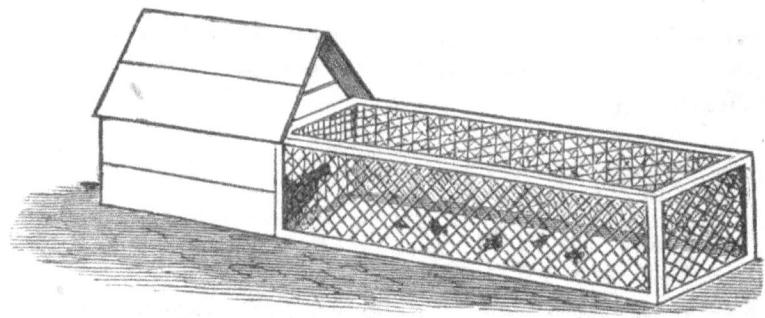

CHICKEN COOP AND WIRE RUN,

feet wide. In such a run all animal depredations may be defied; and in any case we should recommend its use until the chicks are a fortnight old; it saves a world of trouble and anxiety, and prevents the brood wandering and getting over-tired. By having an assortment of such hurdles, portable runs can be constructed in a few minutes of any extent required, and will be found of great advantage until the broods are strong. The hen may also be given her liberty within the prescribed bounds."

DUCK HOUSE.

A plan of a very cheap and pretty duck house is given in the *American Poulterer's Companion*, to be constructed after the style of the engraving herewith given. It should be placed on the bank of a pond or small island

A DUCK TENT HOUSE.

of an ornamental sheet of water. It may be constructed of rough boards thatched with straw, and partly covered with running vines and shrubbery, which would not only be ornamental but make a very pretty and cheap house for aquatic fowls. The interior arrangement of the house may vary according to the means and taste of the proprietor, only providing the ducks with nest boxes, in order that they may lay and incubate undisturbed, and affording proper protection for their young.

TURKEY HOUSE AND NEST.

To save the trouble, says the *Hand-Book of Poultry*, of constantly watching turkeys while they are seeking their nests, there should be a yard inclosing an eighth of an acre for every fifteen birds, where nothing else is

TURKEY HOUSE AND NEST.

allowed to go. Eight feet long pickets, with a white birch or any other brushy bush woven in along the top, will make the most secure inclosure. As early as the first of April nests should be made in this yard.

THE BEST ARRANGEMENT FOR NESTS

are small houses, about three feet by three, gable-shaped, (as shown in engraving,) and three feet high in the center. These should be scattered about the yard, and, if convenient, be partly hidden by an over-covering of brush or something more easily made available. If two or three turkeys incline to one nest, set another house at right angles with that which contains the one they covet, and place several eggs in this new nest, and the probability is, that this will end the trouble; or let them all lay together till one begins to sit, and then shut her in, which will oblige the others to provide for themselves elsewhere.

NESTS FOR LAYING HENS.

The engraving (on page 140,) gives an idea of a wicker-work nest, which is recommended by the *Cottage Gardener* as just the thing for breeders to use, as the hens take to them readily. All that is needed to make them is an auger, a saw, a bill-hook, a clasp knife, a stout piece of leather for hinges, some tacks, a few poles, two inches in diameter, cut fresh from the water willow, some strips, and a few seasoned pieces of boards. Rive the willow rods into laths two-eighths of an inch thick; wattle them on the frame as in the engraving. If water willow cannot be readily obtained, any other wood that is susceptible of weaving into baskets will answer the purpose. This device makes a good cool retreat for hens in hot weather.

TIGHT WOODEN BOX NEST.

It is only necessary the birds should be protected from wind and rain, in order to avoid rheumatism; and this is most effectually done by employing for the nest a tight wooden box, like figure 2, open at the bottom, and also

Fig. 1.

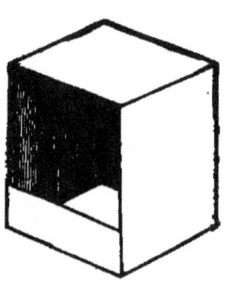

Fig. 2.

in front, with the exception of a strip three inches high to contain the straw. Let one of these boxes be placed in the back corner of the shed, touching the side, the front being turned to the back wall, and about nine inches from it; and the hen will be in the strictest privacy, will be both perfectly sheltered and kept cool, and will never mistake her own nest for the one which may be placed in the other corner. If a third must be made room for, let her nest be placed the same distance from the wall midway between the others, and like them, with the front of the nest to the back of the shed. There will then be still nearly a foot between each two nests for the birds to pass.

CAPONIZING FOWLS.

THE art of caponizing fowls seems to be very little known or understood in this country; we therefore condense from *The Farmers' and Planters' Enclyclopedia of Rural Affairs*, the *modus operandi* as practiced by the best and most experienced English, French and Chinese experts, giving from the same source, also, engravings fully illustrating the subject, together with such information as we have been enabled to gather from other sources. Chickens intended for capons should be of the largest breeds; Brahmas, Cochins, or even Dorkings are fowls well adapted for the purpose.

FOWLS DESIRABLE FOR CAPONS.

Any of the Asiatic fowls are most desirable as they make the best capons; Dorking fowls, however, which are deemed better for the table, answer the purpose very well, as their flesh is so much sweeter and more nutritious than that of almost any breed known. The alteration of a chicken into a capon will, in from ten to twelve months nearly double the size of the bird. Persons wishing to become expert in the operation of making capons would do well to imitate surgeons, who always try their hand upon dead subjects before performing on the living.

MANNER OF PROCEDURE.

The *modus operandi*, however, is quite simple, and in France and Italy is frequently allotted to mere children. Chickens intended for capons may be operated upon at any age, though when between two and three months old is considered much the best time. Old fowls seldom survive the operation. Previous to cutting, the chickens must be kept entirely from food, and even water, for about thirty-six hours, as experiments have determined this time to insure the best chance of success by causing the bowels to be empty, and lessening the tendency to bleeding. The fowl may be secured either in the Chinese mode,— that is to say, lying on its left side with its wings folded back till they meet, and pressed under one foot of the operator, while the other foot is placed on the legs; or it may be held by an assistant, or what adds greatly to the convenience of the operator, especially in relieving him from the necessity of stooping low, the fowl may be confined by straps, etc., to a table, as seen by reference to figures 1 and 2.

HOW TO PLACE THE FOWL.

The fowl being thus secured with its left side downwards, wings clasped behind its back, legs extended backwards, the upper one being drawn the furthest back (see figure 2,) the head and neck being left perfectly free, the feathers are next to be plucked from its right side near the hip joint, in a line between that and the shoulder joint; the space uncovered (*a* figure 2) may be a little over an inch square. Having first drawn the skin of the part backward, so that when left to itself after the operation, it will cover the wound in the flesh, make an incision with the bevel-edged knife (letter *a*,) between the last two ribs, commencing about an inch from the backbone, and extending obliquely downwards about an inch or inch and a half, just going deep enough to separate the ribs, and taking good care not to wound the intestines. A pair of broad blunt hooks, (letters *c, c*,) attached to a piece of elastic whalebone or ratan (*b*) about six inches long are then applied, one hook on each side of the cut, and these being stretched apart by the spring bow, keep the wound open wide enough to give room for the operation.

HOW THE OPERATION IS PERFORMED.

Carefully cut open the skin covering the intestines, which last, if not sufficiently drawn up in consequence of the previous fastening, may be pushed forwards or towards the breastbone, by means of a flat instrument contrived for the purpose, or, what answers equally well, the handle of a teaspoon. When the testicles are exposed to view they will be found to be connected with the back and sides by means of a thin skin which passes over them. This tender covering must be seized with the pincers (*a, a,*) and torn open with the assistance of the sharp edged hook, (*h ;*) after this, with the left hand, introduce the curved spoon under the lower or left testicle (which is generally a little nearer the rump than the right one;) then take the tube (*i,*) and with the right hand pass the loop (*n*) over the small hooked end of the spoon (*h,*) running it down under the spoon and included testicle, so as to bring the loop to act upon the part which fastens the testicle to the back. Then by drawing the ends of the hair-loop backwards and forwards, and at the same time pushing the lower end of the tube towards the rump of the chicken, the cord, or fastening of the testicle, is sawn off. The same process is to be followed with the uppermost or right testicle, after which the separated testicles, together with any blood in the bottom of the wound are to be scooped out with the crooked spoon.

WHEN PERFORMED PROPERLY.

When performed properly, little or no blood of consequence is observed, neither does the bird seem to experience any pain, after the first incision, but will eat food freely if given to it. To enable the operator to produce the sawing movement, the hair or other ligature used, may be tied in a knot, so as to allow the index, or forefinger of the operator's right hand to pass through it. This finger being then turned or rolled repeatedly from side to side, communicates to the loop below the sawing motion which contributes to cut off the testicles. The reason for cutting off the under or lowermost testicle first, is to prevent the blood which may issue from covering the remaining one, thereby rendering it difficult to be seen. After this operation which, if skillfully performed, occupies very few minutes, the hooks are to be taken out, the skin drawn over the wound, and this covered with the feathers plucked off at the commencement of the operation. The chicken is then released, and as soon as let go will take grain or other food eagerly, and in a day or two be restored to its usual health. A person well skilled may operate on fifty chickens without killing more than one or two.

DIFFERENT FORMATION OF FOWLS.

In some fowls the fore part of the thigh covers the last two ribs; in which case care must be taken to draw the fleshy part of the thigh well back, to prevent its being cut, as this might lame the fowl, or even cause its death. For ligatures nothing answers so well as that usually employed by the Chinese, namely, the fiber of the cocoanut husk. This is rough, and makes a loop which saws off and separates the testicle very readily. The next best substance for this purpose is horse-hair. Experiments with fine wire, silk, silk gut, etc., show that these are all inferior to cocoanut fiber and horse-hair.

FOWLS NOT PERFECTLY CAPONIZED.

Sometimes a portion of the testicle adheres and is left behind; in which case the fowls will not prove capons, as will soon be evident, and may be killed for use as soon as the head begins to grow large and get red, and they show a disposition to chase the hens. Then, again, the real capon will make itself known by the head remaining small, the comb and gills losing their bright redness and appearing withered; the feathers of the neck and tail will also grow longer.

AGE TO KEEP CAPONS.

Capons should be kept to the age of fifteen or eighteen months, which will bring them in the spring and summer, when poultry is scarce and bears a high price. Still they should not be killed near molting time, as all poultry then is very inferior. The operation of caponizing fails principally in consequence of the bursting of the skin which incloses the soft matter of the testicles, some of which remains in the bird.

DANGER OF BURSTING OF THE TESTICLES.

Fowls of five or six months old are less liable to have the testicles burst in the operation than younger ones, but they are also more apt to bleed to death than those of from two to four months old. As the large vessel that supplies the entrails with blood passes in the neighborhood of the testicles; there is danger that a young beginner may pierce this with the pointed instrument in taking off the skin of the *lower* testicle, in which case the chicken would die instantly. There are one or two smaller vessels to be avoided, which is very easy, as they are not difficult to be seen. If properly managed no blood ever appears until a testicle is taken off; so that should any appear before that, the operator will know that he has done something wrong. If a chicken die during the operation by bleeding, it is of course as proper for use as if bled to death by having its throat cut.

FOWLS SELDOM DIE AFTER THE OPERATION.

Fowls very seldom die after the operation unless they have received some internal injury, or the flesh of the thigh has been cut through, from not being drawn back from off the last two ribs, where the incision is made; all of which accidents may be liable to occur with young practitioners.

TESTICLES WHEN FOUND LARGE.

When the testicles are found very large, the silver tube may be too small for the operation; in this case a larger one made of small bamboo or elder, about three-eighths of an inch in diameter, may be substituted, with a strong cocoanut string or ligature. But for chickens of small and medium sizes, the silver tube, with a horse-hair in it, will answer perfectly well.

MARKING CAPONS BEFORE LETTING THEM RUN.

When a chicken has been cut it is necessary before letting it run to put a permanent mark upon it; otherwise it would be impossible to distinguish it at first from others not operated on. Cutting off the outside or inside toe of the left foot, will enable one to distinguish them at a distance. Another mode is to cut off the comb, then shave off the spurs close to the leg, and stick them upon the bleeding head, where they will grow and become ornamental in the shape of a pair of horns. This last mode is perhaps the best, but it is not so simple and ready as the first. Whichever plan is adopted, the fowl should be marked before performing the operation.

TREATMENT OF WINDY SWELLING IN CAPONS.

It is very common, after the operation, and while the wound is healing, for the side to puff out with a windy swelling. This may be relieved by making a small incision or puncture in the skin, which will let the wind escape. Those fowls make the best and finest capons which are hatched early in the spring; as they can be cut before the hot weather comes, which is a great advantage. The operator should not be discouraged with the first difficulties; for with practice they will disappear; every year's experience will render one more expert, until the cutting of a dozen or more fowls before breakfast will be a small matter.

DISSATISFACTION WITH THE OPERATION.

It may be well to give a warning against becoming dissatisfied with the instruments. A raw hand, when he meets with difficulties, is apt to think the tools are in fault, and sets about to improve them and invent others; but it may be only himself that lacks skill, which practice alone can give. Those who have devoted much time and attention to the subject say that they have found the old Chinese instruments, illustrations of which are herewith given, preferable to all others. In addition to these instruments, a regular Chinese set contains a flat kind of spatula something like the upper part of a spoon handle. This is about four inches long and half an inch wide, and slightly curved at each end in opposite directions. It is for the purpose of pushing the intestines out of the way, an office very well performed by the handle of a teaspoon. The engravings given below represent the instruments used in making capons, according to the Chinese method, reduced only about one-fourth their actual sizes.

DESCRIPTION OF THE IMPLEMENTS TO BE USED.

a, a knife, the edge of which resembles that of a chisel with a bevel or slanting edge, half an inch in the greatest width; the other end or handle consists of two forcep blades terminating at *a*, *a*, in slender points, and forming spring forceps. The whole length from the cutting edge to the end of

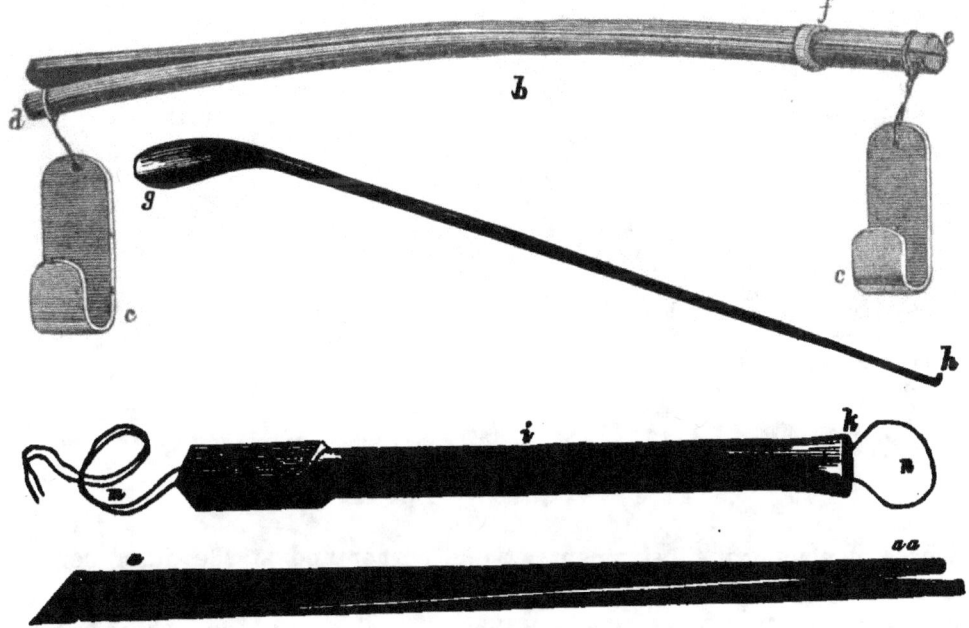

DESCRIPTION OF IMPLEMENTS USED.

the pliers is about six inches. *c, c*, two broad hooks of silver or other metal, each half an inch in width and one and a half in length. *b*, an elastic bow, six inches long, made of whalebone or ratan, about the thickness of a large quill, and split horizontally into two pieces. To the ends of this bow the broad hooks are attached by strong cords about half an inch long. At the end *d*, the cord embraces only the lower half of the split bow, whilst both pieces are included in the string, at the end *e*. *f*, is a small ring which encircles both portions of the bow. When the hooks are first put in and only half the strength of the bow is required to act upon them, this ring is slipped to the end *e*. But if the whole strength of the bow is needed to force the hooks apart and stretch the wound open, the ring is passed towards the end *d*. Thus, by means of the split bow and sliding ring, the strain upon the hooks can be increased or slackened at pleasure. *i*, a tube of silver or other metal three or four inches long, made square at the upper, and flattened at the lower end *k*, to the width of three-tenths of an inch; this tube is for the purpose of passing the fiber or hair ligature *m*, forming the loop *n*. *g*, a narrow curved spoon, the slender handle of which tapers off, and has a steel point fitted into it, furnished at the extremity with a very small hook, *h*; the inner edge of this hook is sometimes sharpened.

THE OPERATING TABLE.

The operating table is represented in the following cut, figure 1. This table may be about two and a half feet long by one and a half feet wide, and two and a half feet high. At two of its corners it can have a raised molding about half an inch high, extending along the sides six or nine inches, for the

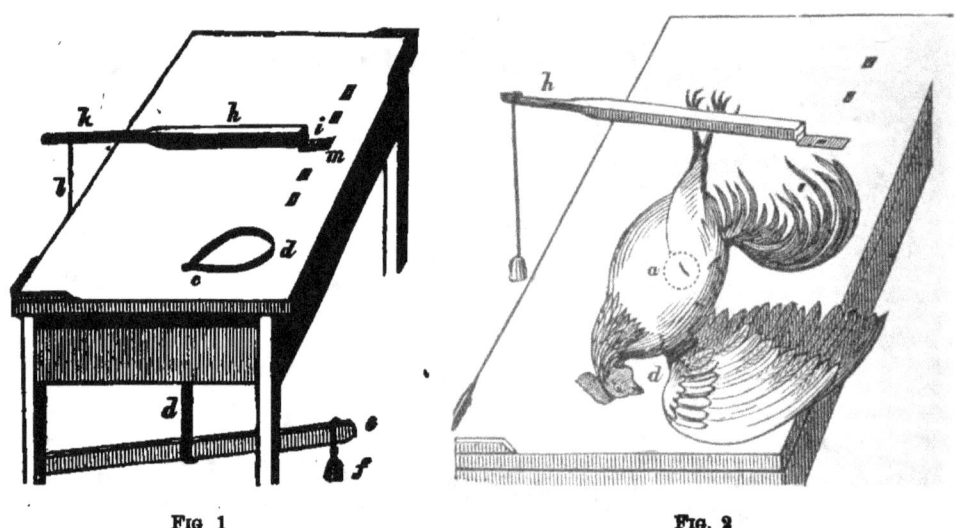

FIG 1   FIG. 2

purpose of placing the instruments at one corner and at the other some of the feathers under a stone, to keep them from being blown away. On one side there is a slit *c*, passing through the table, about one and three-quarters of an inch long by one-half an inch wide, running diagonally; being about three inches from the end and six and a half from the side. Through this slit the padded band or soft list, *d, d*, for confining the wings, passes below to be attached to the lever *e*. This lever has a four or five pound weight hung to it, and works on a screw or pin, by which it is attached to the leg. When not in use the lever rests on a pin or ledge in the other leg. On being let down, the attached band clasps the wings of the chicken lying on the table, with greater or less force as the weight is drawn to or from the end of the lever. The next thing to be described is the lever, *h*, upon the table, the object of which is to hold down the legs as these are extended backwards. This lever is padded beneath, and is furnished with a hinge at *i*, which admits of being raised at the end *k*, it projects beyond the edge of the table, and has also a five pound weight suspended by the string *l*, which increases or diminishes the pressure by being moved to or from the table. Through one portion of the hinge an iron screw, *m*, passes beneath the table where the end is secured by a nut. This screw or pin allows the lever to move sidewise, whilst the hinge admits of its being raised or let down. A range of holes, about one-third of an inch wide, is made through the table, to receive the pin of the lever, as this has to be placed nearer to or farther from the slit *c*, according to the size of the chicken. The first hole is about eleven inches from the nearest end; the second, fourteen inches; the third, seventeen inches. The last is adapted to very large cocks or even turkeys.

POSITION OF THE FOWL ON THE TABLE.

In figure 2, the position of the fowl when secured, lying upon its left side upon the table, is represented, *d* being the wing-band, *h* the lever placed over the legs, and *a* the place where the incision is made. The table is a refinement in the art of caponing which we believe is altogether new, notwithstanding the thousands of years which have elapsed since the operation has been habitually practiced. The difficulty of making a subject, apparently simple, well understood by persons to whom it is entirely new, is, we think, a sufficient apology for the length of the details given.

USEFULNESS OF CAPONS.

In France and other countries, besides furnishing a luxurious food, capons are made useful in taking care of broods of young chickens, ducklings, turkeys, and pheasants, which they are said to do much better than hens, owing to their larger size and thicker coats of feathers. The moment the chickens are hatched they are taken from the hens and given to a capon, who rears them with all the care of a parent, often having a small bell attached to his neck, the tinkling of which serves the purpose of keeping the brood about him, similar to the clucking and maternal sounds of the mother. Should he show a disposition to treat the young chickens roughly at first, he may be confined alone for a day or two in a dark place, after which if they be put with him he will be pleased with their company and continue to take care of them. The hen is cooped, and well fed until she regains the flesh and strength lost whilst sitting, and then turned out to lay again. In this way the poulterer is enabled to raise a large number of chickens from a few hens. The capon generally brings double or treble the price of common poultry.

ANOTHER MODE OF PLACING THE FOWL.

Figure 3 shows a different mode of preparing the fowls for caponizing, which only requires a very little trouble to make. In the first place you confine the cockerel between the two weights, on a table or board, as you choose, (see engraving,) laying him with the left side downwards, and placing his wings locked across the back, which assists in holding him down; the legs extended backward, with the upper one drawn furthest out. Leave the head

Fig. 3.

and neck free. Pluck the feathers from the right side, near the hip-joint, from an inch to an inch and a half in diameter, and on a line with the shoulder. Then proceed with the business as directed elsewhere.

TO CAPONIZE YOUNG PULLETS.

Young pullets may also be caponized, so as to deprive them of their reproductive powers. It has the same effect upon them that it does upon the cockerels — rendering them more easy to fatten. A pullet that has no inclination to lay regularly can be got rid of in this way with profit to the breeder. The usual method of making *poulardes*, as caponized hens are termed in France, is to extirpate the egg-cluster, or *ovaries*, in a similar manner to extracting the testicles from young cocks. Mr. YARRELL says, however, "that it is quite sufficient merely to cut across the egg-tube or oviduct, with a sharp knife." Birds after once being caponized are never subject to the natural process of molting.

ANATOMY OF THE EGG.

THE OVARIUM.

In a laying hen, M. VIELE, an eminent anatomist of France, says, may be found, on opening the body, what is termed the *ovarium*—a cluster of rudimental eggs, of different sizes, from very minute points up to shapes of easily-distinguished forms. These rudimental eggs have as yet no shell or white, these being exhibited in a different stage of development; but consist wholly of yolk, on the surface of which the germ of the future chicken lies. The yolk and the germ are enveloped by a very thin membrane. When

THE RUDIMENTAL EGG,

still attached to the *ovarium*, becomes longer and larger, and arrives at a certain size, either its own weight, or some other efficient cause, detaches it from the cluster, and makes it fall into a sort of funnel, leading to a pipe, which is called the *oviduct*. Here

THE YOLK OF THE RUDIMENTAL EGG,

hitherto imperfectly formed, puts on its mature appearance of a thick yellow fluid; while the rudimental chick or embryo, lying on the surface opposite to that by which it had been attached to the *ovarium*, is white, and somewhat like paste. The white, or

ALBUMEN OF THE EGG,

now becomes diffused around the yolk, being secreted from the blood vessels of the egg-pipe, or *oviduct*, in the form of a thin, glassy fluid; this is prevented from mixing with the yolk and the embryo chicken by the thin membrane which surrounded them before they were detached from the egg-cluster, while it is strengthened by a second and stronger membrane, formed around the first, immediately after falling into the oviduct. This second membrane, enveloping the yolk of the germ of the chicken, is thickest at the two ends, having what is termed bulgings by some, and *chalazes* by anatomists; these bulgings of the second membrane pass quite through the white at the ends, and being thus, as it were, embedded in the white, they keep the inclosed yolk and germ somewhat in a fixed position, preventing them from rolling about within the egg when it is moved. The white of the egg being thus formed, a third membrane, or, rather, a double membrane, much stronger than either of the first two, is formed around it, becoming attached to the chalazes of the second membrane, and tending still more to keep all the parts in their relative positions.

PROGRESS OF THE FORMATIONS.

During the progress of these several formations, the egg gradually advances about half way along the oviduct. It is still, however, destitute of the shell, which begins to be formed by a process simliar to the formation of the shell of a snail, as soon as the outer layer of the third membrane has been completed. When the shell is fully formed, the egg continues to advance along the oviduct, till the hen goes to her nest and lays it. From ill-health, or accidents, eggs are sometimes excluded from the oviducts before the shell has begun to be formed, and in this state they are called wind eggs.

THE EGG HAS SIX DIFFERENT ENVELOPES.

Reckoning, then, from the shell inward, there are six different envelopes, of which one only could be detected before the descent of the egg into the oviduct,—the shell; the external layer of the membrane lining the shell; the internal layer of same lining; the white, composed of a thinner liquid on the outside, and a thicker and more yellowish liquid on the inside; the bulgings, or chalaziferous membrane; and the proper membrane. One important part of the egg is

THE AIR-BAG,

placed at the larger end, between the shell and its lining membrane. This is about the size of the eye of a small bird in new-laid eggs, but is increased as much as ten or twelve times in the process of hatching. The air-bag is of such great importance to the development of the chicken— probably by supplying it with a limited atmosphere of oxygen—that, if the blunt end of an egg be pierced with the point of the smallest needle, the egg cannot be hatched.

DOUBLE-YOLKED EGGS.

Instead of one rudimental egg falling from the ovarium, two may be detected, and will, of course, be inclosed in the same shell, when the egg will be double-yolked. The eggs of a goose have, in some instances, been known to contain even three yolks. If the double-yolked eggs be hatched, they will rarely produce two separate chickens, but, more commonly, monstrosities—chickens with two heads, four legs, and the like.

THE SHELL OF THE EGG.

The shell of the egg, chemically speaking, consists chiefly of carbonate of lime, similar to chalk, with a small quantity of phosphate of lime and animal mucus. When burned, the animal matter and the carbonic acid gas of the carbonate of lime are separated; the first being reduced to ashes, or animal charcoal, while the second is dissipated, leaving the decarbonized lime mixed with a little phosphate of lime.

THE WHITE OF THE EGG.

The white of the egg is without taste or smell, of a viscid, glairy consistence, readily dissolving in water, coagulable by acids, by spirits of wine, and by a temperature of one hundred and sixty-five degrees Fahrenheit. If it has once been coagulated, it is no longer soluble in either cold or hot water, and acquires a slight insipid taste. It is composed of eighty parts of water, fifteen and a half parts albumen, and four and a half parts mucus; besides giving traces of soda, benzoic acid, and sulphureted hydrogen gas. The latter, on an egg being eaten from a silver spoon, stains the spoon a blackish purple, by combining with the silver, and forming sulphuret of silver. The white of the egg is a very feeble conductor of heat, retarding its escape, and preventing its entrance to the yolk; a providential contrivance not merely to prevent speedy fermentation and corruption, but to arrest the fatal chills, which might occur in hatching, when the mother hen leaves her eggs, from time to time, in search of food. Eels and other fish which can live long out of water, secrete a similar viscid substance on the surface of their bodies, furnished to them, doubtless, for the same purpose.

THE YOLK OF THE EGG.

The yolk has an insipid, bland, oily taste; and, when agitated with water, forms a milky emulsion. If it is long boiled it becomes a granular, friable solid, yielding, upon expression, a yellow, insipid, fixed oil. It consists, chemically, of water, oil, albumen and gelatine. In proportion to the quantity of albumen, the egg boils hard.

THE WEIGHT OF EGGS.

The weight of the eggs of the domestic fowl varies materially; in some breeds averaging thirty-three ounces per dozen, while in others, but fourteen and a-half ounces. A fair average weight for a dozen is twenty-two and a-half ounces. Yellow, mahogany and salmon-colored eggs are generally richer than white ones, containing, as they do, a large quantity of yolk. These are generally preferred for culinary purposes; while the latter, containing an excess of albumen, are preferred for boiling, etc., for the table.

FACTS ABOUT INCUBATION.

We are informed by M. Tegetmeier that, in breaking a number of eggs into a basin, a small circular speck may be observed upon each yolk. This speck is the rudiment of the young chick, and the construction of the egg is such that, on whatever side it is turned, the rudimentary germ is always uppermost, so as to receive the heat from the sitting hen. The mechanism by which this is managed is very simple:—The lower side of the yolk is weighed or ballasted by two heavy twisted masses of very firm albumen, which keep the germ constantly uppermost. Contrary to general belief, these ballasting weights are found in all eggs, whether laid by pullets or old hens. If an egg has been set upon for even a few hours, the size of the germ is increased, and if left in the nest of a sitting hen for twenty-four hours, small blood vessels may be seen forming a beautiful zone around it. The yolk, like the white, is composed of concentric layers, which may be seen when it is boiled hard, and from the germ a tube runs to a central hollow or cavity, often noticeable when an egg boiled hard for salad is cut across.

WHEN A FECUNDATED EGG IS PLACED UNDER A HEN,

or deposited in an incubator, and subjected to a temperature somewhat above one hundred degrees, the germ undergoes a remarkable series of alterations, being gradually developed into the perfect chick. During the period of incubation, various changes occur. The air-vesicle at the end gradually becomes larger in proportion as the water of the albumen evaporates, through the pores of the shell. During its development, the chicken derives its nourishment chiefly from the yolk; and shortly before birth the remainder of the yolk is drawn into the abdomen, and passing into the digestive canal, constitutes the first food of the newly hatched animal. During incubation, the blood of the chick is aerated by passing through a series of vessels in a temporary respiratory membrane which lines the porous shell; this makes its appearance on the third day, and gives rise to that opacity of the fertile egg which may always be observed. It is not until the nineteenth day of incubation that the beak of the chick ruptures the enlarged air vesicle, and it then only commences to breathe by means of its lungs. This is accompanied by a peculiar sound known as "tapping," which is merely respiratory, and is not caused by contact of any kind between the beak of the chick and the interior of the shell.

INCUBATORS.

THE HATCHING AND REARING OF CHICKENS BY ARTIFICIAL MEANS
has not received that attention which its importance demands. The business of raising poultry in this country has been very limited and its operations very primitive. It is only for a few years past that farmers and fanciers have taken hold of it with any degree of earnestness; and for the very short time they have given it their attention their success has been wonderful, and plainly shows what can be done by a little attention and perseverance. Poultry and eggs should be one of the staple articles of subsistence for the people, where now only a few, comparatively, share in these luxuries, on account of the high prices these necessities of life command. Chickens should never sell for over twelve cents per pound, and eggs twelve cents per dozen; and the present prices could be reduced to the above standards by means of the artificial methods of hatching and raising them. If each individual who takes an interest in

RAISING POULTRY EITHER FOR PLEASURE OR PROFIT
had the means of hatching out only one hundred and fifty chickens every three weeks it would quintuple the stock of the country. And what a saving of time and labor; and especially when such an instrument could be managed by the younger folks. An incubator that would hold two hundred eggs would produce in six sittings, on an average, nine hundred chickens, even allowing for the loss of fifty eggs by various means at each sitting and would perform the work of sixteen hens every three weeks; and the hens could be brought back to the business of laying again in a very short time. Of course we do not mean to keep the hens from having one good sitting a year, which is so necessary for their rest, comfort and health. We have not in this estimate taken into account the length of time it takes the hen to raise her brood so that they may look out for themselves. This will be treated of more fully under the title of "Artificial Mothers."

THE DIFFERENT INCUBATORS.

From the number of successful experiments that have been made by different inventors and scientific men, we are more fully convinced of the practicability of the plan, and that it wants only encouragement from the people to make it a success.

THE EGYPTIAN MODE OF HATCHING EGGS.

The Egyptians hatched eggs in ovens on an immense scale with great success. Thousands of thousands of chickens were hatched in a season in this way. REAUMUR succeeded in

HATCHING CHICKENS IN WOODEN CASKS

by surrounding them with fresh manure in a state of fermentation, and this method, though not the most pleasant, is still employed in France with good results.

CANTELO was successful in supplying the heat from above in imitation of the hen. The elaborate contrivance of MINASI was a very ingenious and successful effort. It could hold two hundred eggs. The chickens were all strong, healthy and vigorous, but the great drawback to these two methods of CANTELO and MINASI was the expensiveness.

GEYELIN'S INCUBATOR

is one which we do not think very desirable at the present day; still we give, in this connection, a description of it, with illustrations, more for the purpose

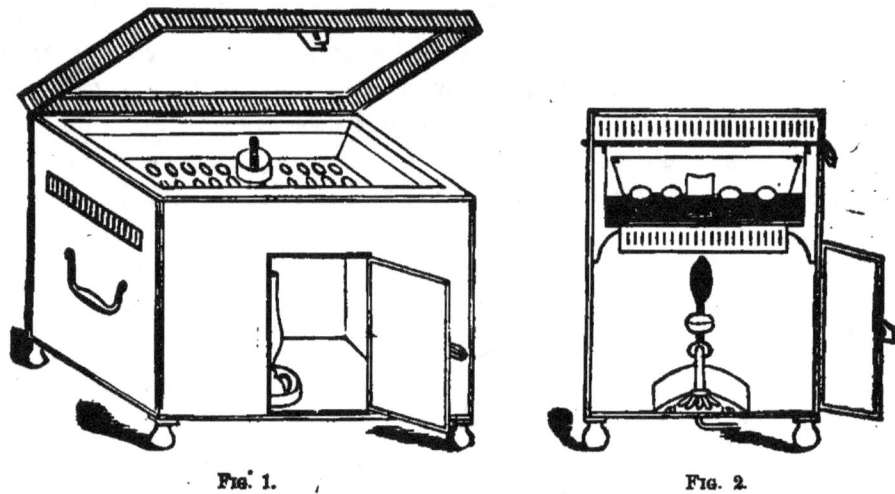

FIG. 1.   FIG. 2.

of showing the marked improvement made in these machines than for any other reason. Figure 1 represents a perspective elevation of an artificial pen for hatching; figure 2, a transverse section of the same. The hatching apparatus consists of separate parts: first, a glass-covered box; second, a water tank; third, a floating vessel; fourth, a gas or oil lamp.

The glass-covered box is made of japanned tin; it has a glass door through which the light can be seen; the bottom of this box is perforated in the center for the admission of air to the lamp, and the other part is carpeted to receive the chickens as they leave their shells. About twelve inches from the bottom are four brackets, to receive the water tank; the lid has a perforated border for the escape of the vitiated air and steam from the water. The sides are provided with handles for carrying the box from one place to another, and it stands on four knobs to allow a free passage of air underneath. The water tank is made of tin, and a little smaller than the box, so as to allow half an inch free passage of air all round. The floating vessel is made of tin, and is a trifle smaller than the water tank, so as to allow of its floating in it. The center of this vessel has an oval opening, in which a registering thermometer is kept to show at all times the temperature of the water. The bottom of this vessel is covered about one inch deep with silver sand, on which the eggs are placed. By means of the central opening, and that between the tank, the temperature is kept in a constantly moist state. The lamp can be for oil or gas, but gas is certainly preferable. The management of the apparatus is so simple that it can be attended to by a child, and only a very few directions will be necessary:—1. Fill the tank with hot water till the floating vessel reaches the top level, then see that the water has a temperature of about one hundred and twelve degrees, after which light the lamp, and should the heat of the water increase, reduce the flame; but if the temperature rises or decreases but slowly, it can be regulated by admitting more or less air through the door of the box. 2. The principal point, however, is, that the temperature on the sand should not vary much from one hundred and five degrees, and it will be found that with water-heat of one hundred and twelve degrees, the sand will be one hundred and five, and on the eggs ninety-eight degrees. For beginners, however, it is always best to put the apparatus in action a day or two before placing eggs in it. 3. Turn the eggs once or twice a day, and keep the water replenished as it evaporates.

The only incubators that are considered at all practicable are those of M. CARBONNIER, Mr. BRINDLEY and Mr. F. SCHRODER.

M. CARBONNIER'S INCUBATOR

was quite a simply constructed machine. The heating apparatus consisted of a tin or copper cistern or boiler of any desired size made with a flat bottom and heated by a lamp, for which a chamber was provided in one end. The lamp was so constructed as to burn for a certain length of time without attention, and it was essential that the lamp chamber should be in the end of the cistern that there might be a regular circulation of the water. The cistern was kept nearly filled; and the constant immersion of the thermometer was necessary to show the temperature. The eggs were placed in a drawer under the cistern on a little hay. They were not exposed to the direct heat of the cistern, but were covered with a piece of canvas, on which is spread a layer of sawdust half an inch thick. The sawdust became warmed by the heat of the cistern and resting gently upon the eggs warms them in a more natural way than any preceeding incubator we know of. In the egg drawer should be a second thermometer to indicate the heat the eggs were subjected to. The temperature of the sawdust may be kept at a standard of one hundred and two or one hundred and three degrees Fahrenheit, and regular attention was necessary to insure this. The eggs were withdrawn every day and exposed to the cold air for about twenty minutes, and turned over as often, and the sawdust laid again upon them, and sprinkled with water heated **to one hundred and five degrees**, so as to make it slightly moist.

THE ARRANGEMENT OF MR. BRINDLEY'S INCUBATOR

is shown by figure 3. It is a copper boiler heated by a lamp or **gas jet, B**, furnished with a reservoir, also marked B, carefully constructed **to burn with steadiness**. From this boiler the hot water flows constantly through a sys-

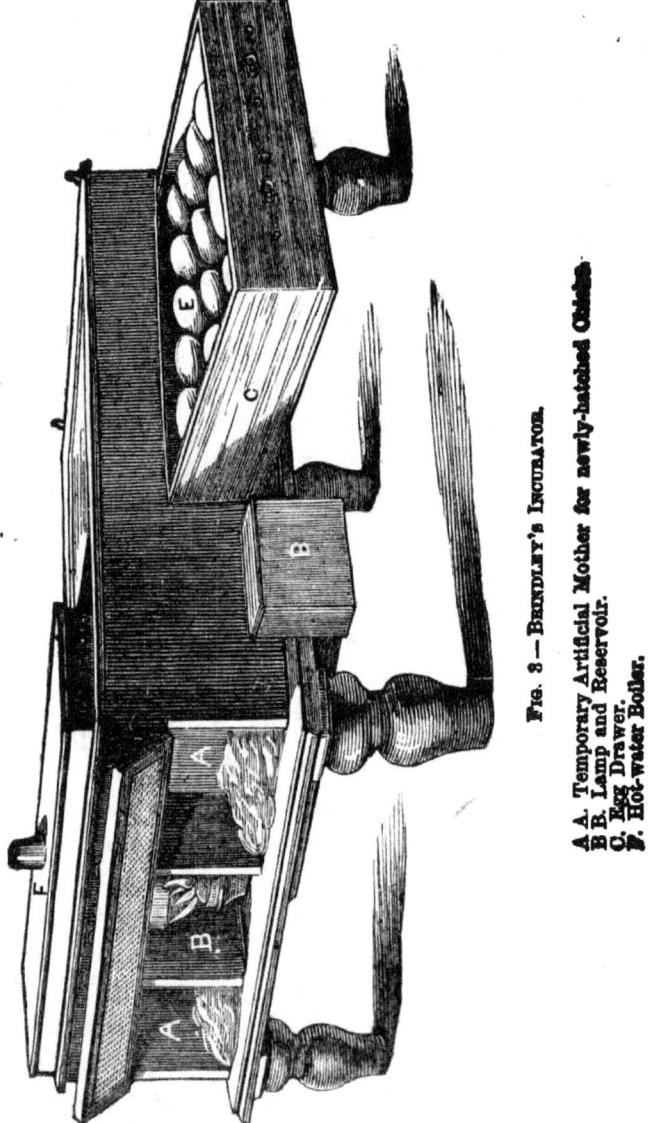

FIG. 3.—BRINDLEY'S INCUBATOR.
A A. Temporary Artificial Mother for newly-hatched Chicks.
B B. Lamp and Reservoir.
C. Egg Drawer.
F. Hot-water Boiler.

tem of metal pipes, arranged in a horizontal place between two plates of glass, which thus forms a *hot-air chamber* heated by the pipes. Under the lower glass plates slides the drawer C, lined with felt which contains the eggs. At each side of the lamps at A are

TEMPORARY RECEPTACLES, OR ARTIFICIAL MOTHERS,

to receive the chickens for the first day, after which they may **be removed** and provided for separately. The hot air chamber is provided with a "safety valve," acted on by the expansion of mercury, which opens at a given temperature. This valve seems to have been employed first by M. VALLEE of the Jardin des Plantes, Paris. Mr. BRINDLEY's valve seems to have been superior to all those shown before him, and to answer all reasonable purposes. Mr. WRIGHT thinks it impossible to make any valve the *sole regulator*, and expect it to keep the heat uniform. He is under the impression that when the heat becomes two or three degrees too high all that is expected is, that the valve will open and admit cold air to reduce the temperature; but if the air is really hot the valve, though open, cannot entirely keep the heat down, nor can it guard against a lower temperature than is proper. We shall show hereafter how the heat is regulated by the Graves' plan, so as to obviate these difficulties.

BRINDLEY's machine differs radically in *principle* from the preceding one, as also from Mr. F. H. SCHRODER's, in that the valve is not employed directly to warm the eggs but simply to impart heat to a chamber of hot *air* through which the heat is communicated. In other respects the management is similar. The eggs require to be withdrawn and cooled once a day; should be carefully turned and sprinkled with warm water, which should also be allowed to moisten the felt lining of the tray in which they are contained.

THE INCUBATOR OF MR. F. H. SCHRODER

is shown in figure 4. He has adopted an altogether distinct and separate boiler, which is not shown; but which is connected with the hot water tank

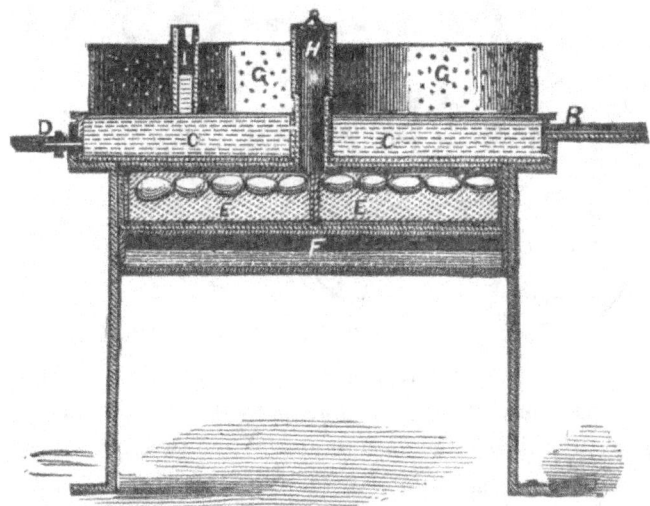

FIG. 4.

C, of the incubator by two pipes; B being the inlet pipe, and D the outlet. This tank is provided with an open table, I, in which a thermometer can be placed to show the temperature, and with a ventilating tube H, which is open at top and bottom. Under the tank slide the egg drawers E, which in area resembles the quadrant of a circle — this is of a circular form.

THE INCUBATOR BEING OF A CIRCULAR FORM,

the bottom of these drawers are of perforated zinc, and partly filled with sand, both to preserve the heat and to form a convenient and warm receptacle for the newly-hatched chickens. Curtains are provided to surround the sides of the incubator, and thus guard, in some measure, against change of temperature in the apartment. In using this incubator the egg-drawers E are partly filled with chaff or other similar material, on which the eggs are deposited. The water from the cold water cistern F, underneath them, slowly evaporates with the heat above, and preserves a gentle moisture around the eggs during the process of incubation, percolating as it does through the chaff and perforated bottom of the egg-drawer; ventilation takes place through the middle shape or pipe H. Sprinkling the eggs is not necessary in Mr. SCHRODER'S plan; all that you need to do is to replenish the cold water tank F when exhausted. The eggs, however, as in all incubators, should be withdrawn, cooled half an hour and turned every day.

THE INCUBATOR OF COL. STUART WORTLEY,

represented in figure 5, is described by Mr. WRIGHT in his work as superior to all the rest; but, at the same time, admits that it has not yet been generally tested. D, is a saddle-backed or other convenient boiler, furnished

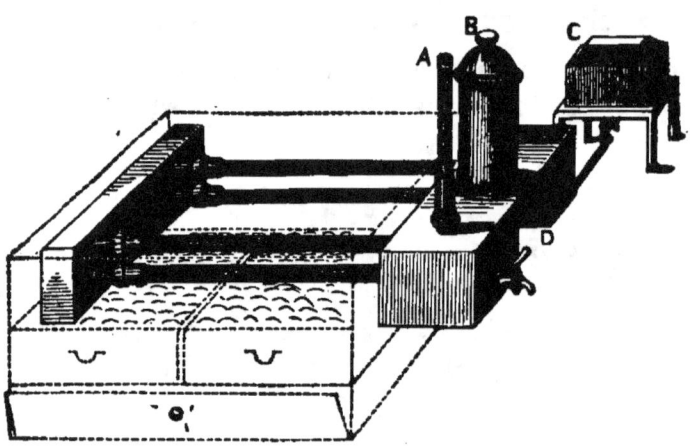

FIG. 5 — COL. STUART WORTLEY'S INCUBATOR.

with a steam dome, by which the steam is collected and allowed to escape. Connected with the boiler is a simple cistern C, by which the hight of the water is always kept uniform, a glass gauge, A, showing the hight at a glance. The water in the boiler is always kept boiling, and circulates therefore at a uniform temperature through the pipes, which heats the egg chamber. These pipes pass through padded holes, and hence by sliding them in more, there is greater heat imparted for cold weather, or by withdrawing them a little the temperature will fall.

THE GREAT IMPROVEMENT OF COL. WORTLEY'S INCUBATOR

is the control he has over the variations of temperature. He seems to take advantage of the natural law, which, without trouble, gives him always a temperature of one hundred and twelve degrees, and then provides for changes by giving more or less of heating surface.

THE AMERICAN INCUBATOR.

This incubator, represented by the cuts; figure 4, representing the outward appearance of the machine, and figure 5 the inside arrangements, was awarded the first premium at the Pennsylvania State Poultry Exhibition, held in the city of Philadelphia.

C, figure 5, is the nursery for young chickens for the first week after hatching, D being a ventilator, of which there is a corresponding one in the rear. B, B, B, B, is the boiler, by which the heat is generated by means of a lamp L. N, N, N, N, are the nests or drawers for the eggs. The two lower

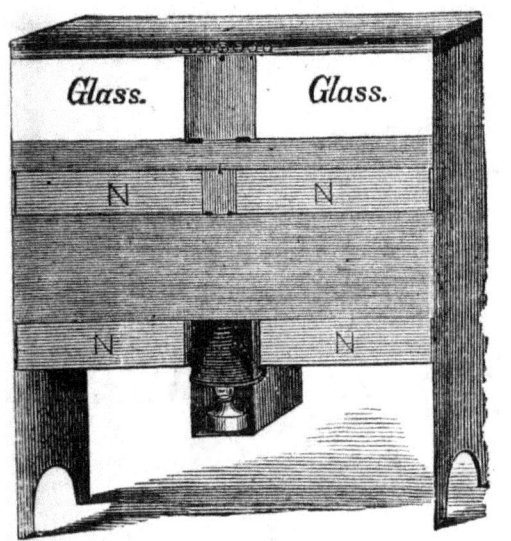
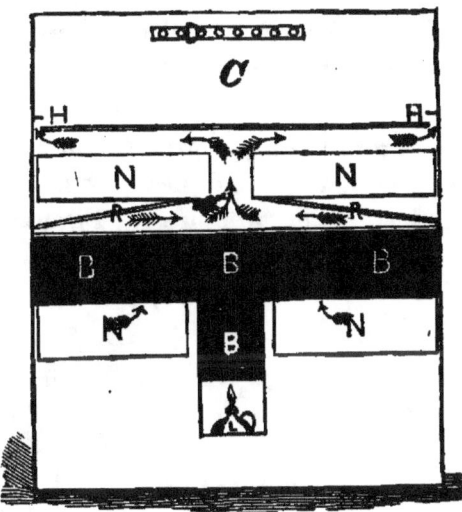

Fig. 4.   Fig. 5.
THE AMERICAN INCUBATOR.

ones being directly under the boiler, the heat is applied above the eggs, the same as in natural incubation. To apply the heat in the same manner to the eggs in the upper drawers, the inclined shelves, R, R, are used, (the two drawers being tight-bottomed also.) The heat is thus forced to ascend as shown by the arrows, and passes over the eggs, escaping by the ventilators at H, H, into the nursery, C, where it is again utilized for the young chicks. V, V, are tubes going through the boiler, serving for stays to keep the boiler from bulging or collapsing, and also answering for ventilating the lower tier of drawers. The boiler is so constructed as to keep the water in constant circulation, thus securing, as is claimed, a uniform heat in all portions of the boiler, with a smaller consumption of fuel than by any other method.

The great trouble in hatching machines heretofore presented to the public, has been the impossibility of keeping the drawers below the boiler at the same temperature as those above. It is claimed that in this machine that difficulty is entirely overcome; and that by the arrangement of the ventilators heat is more perfectly under control than has been before attained. This—an even temperature—is the most important point to be secured; with it success is almost certain; without it, almost impossible. The proper heat is one hundred and three degrees Fahrenheit; the minimum being one hundred degrees, and the maximum one hundred and five degrees. It is not always fatal to let the heat go below one hundred degrees, if not allowed to remain so any great length of time; but a heat of one hundred and seven or one hundred and eight degrees is almost certain death to all unhatched chickens. The period of incubation is not shortened, as many suppose; or, at most, only one day—twenty days being the average time; and we frequently see hens bring off their broods in that time, if close sitters. The chicks come out remarkably strong and healthy, and are always free from vermin; and after the first few days require no more care than if hatched under a hen.

THE GRAVES' INCUBATOR.

Figure 1 is a perspective view of a portion of an incubator case, showing the ventilating and heat-regulating devices. Figure 2 is a transverse, vertical section of figure 1. Figures 3 and 4 are views in detail of the heat-

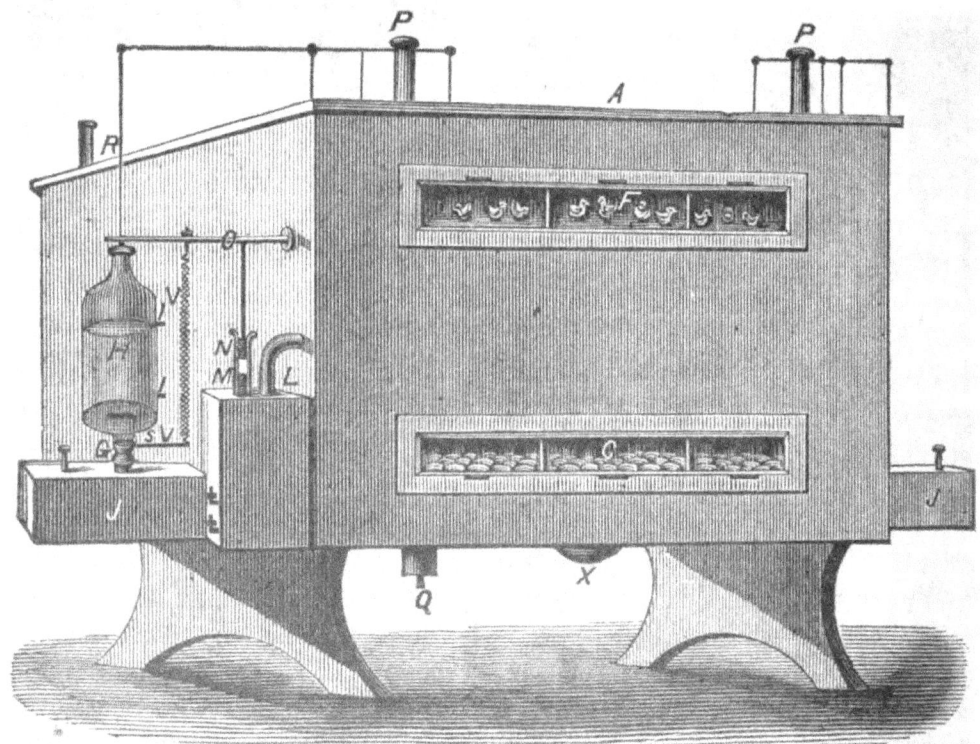

FIG. 1—PERSPECTIVE VIEW OF GRAVES' INCUBATOR.

regulating devices; and figure 5 (shown elsewhere,) is a view of an artificial mother, or protector. The object of this invention is to maintain a definite degree of temperature in incubators heated by water, and it consists, mainly, of two horizontal glass tubes, closed at one end, containing alcohol, and located under the hot water tank, each tube communicating with a vertical cylinder filled with mercury, one at each end of the incubator; in which cylinders are cork pistons or floats, having rods attached to pivoted levers, which are so connected with the regulators on the heating lamps and ventilating valves, communicating with the incubating chamber, that the rising of said floats or pistons beyond a certain point by the expansion of the alcohol will act to check the flames of the lamps and open the ventilating valves, thus decreasing the temperature of the air and water, while the depression of said floats, in consequence of the contraction of the alcohol, will produce an opposite effect and highten the temperature, the parts being so arranged as not to be affected by the medium temperature, at which the incubator is to be kept, but only by higher or lower degrees.

EXPLANATION OF THE ILLUSTRATION.

In the drawing A represents the incubator, which is divided into several compartments, as shown in figure 2, viz.:—The cold water tank, B; incubating space, C; hot water tank, D; protecting or heat-retaining space, E; and drying loft, F. The ends of the incubator are provided with lamps, G, which heat the water in reservoirs, H. These latter communicate, through tubes, I, I, with the hot water tank, D. J, is a reservoir, which

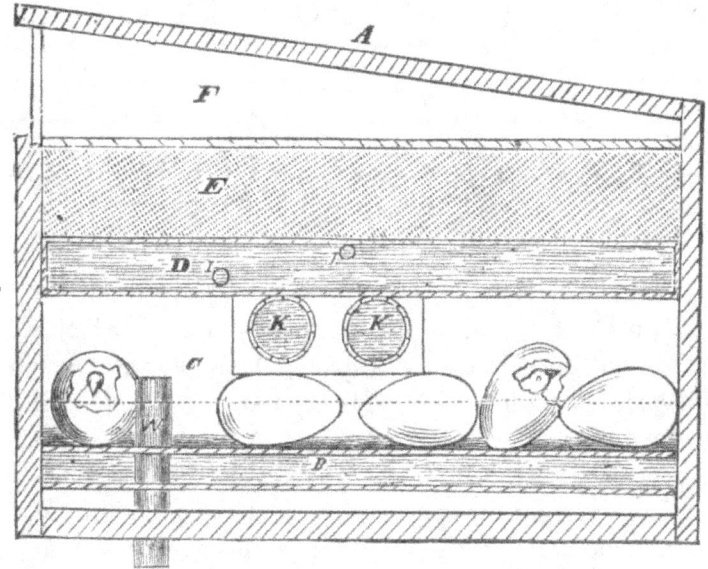

Fig. 2.—Transverse Section of Graves' Incubator.

supplies oil to the lamps. K, K, represent glass tubes under tank, D, and in contact with the bottom thereof. Said tubes are filled with alcohol, or other expansible fluid, and communicate at their outer ends through the bent tubes, L, with the cylinders, M, which contain mercury. N, represents a piston rod, attached to a cork piston, or float, N, in the cylinder, M. The upper end of rod, N, is attached to an arm or lever, O, which is pivoted at one end and swings freely at the other. P, represents a valve, which communicates with the incubating space, C, and is connected by wires, R, with the free end of lever, O; said wires are not rigidly connected with valve, P, but have a sliding attachment. S, represents the lamp-burner, which is provided with the tube, S, which is beveled off at one side, as shown. T, is a guard or regulator, which is journaled on shaft, t, beside the tube, S, and when not in operation inclines from the same. The shaft, t, is bent on the outside of the burner into an elbow or crank, U, which is connected by the spiral spring, V, to the lever, O. The operation of this invention is as follows:—The standard temperature for hatching eggs is about one hundred and two degrees Fahrenheit, at which point this device is arranged to remain inoperative; but, when the water in tank, D, becomes heated above this point, the expansion of the alcohol in tubes, K, causes the cork float or piston, N, to elevate the rod, N, and lever, O, which latter being connected to valve, P, by wire, R, and to regulator, I, by spring, V, opens valve, P, and causes regulator, T, to close over the beveled side of tube, S, thereby lowering the flame in proportion to the nearness it approaches the tube.

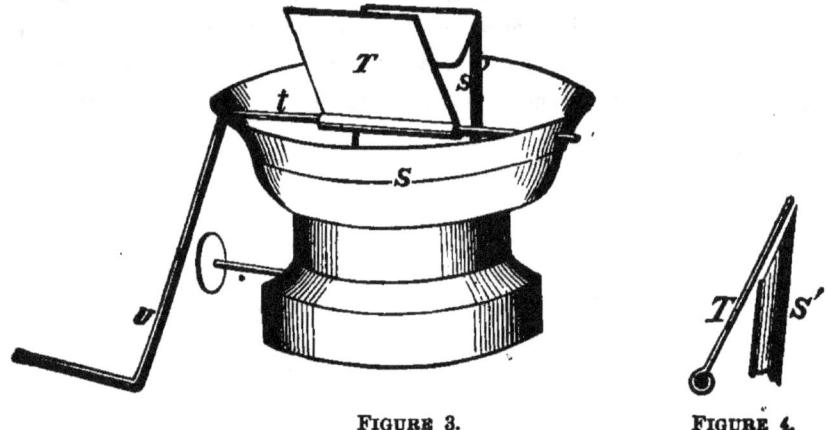

FIGURE 3.   FIGURE 4.

When the valve, P, is opened, warm air rushes up through tube, W, and out through said valve, thus cooling the space, C; while the flame of the lamps, being diminished, the temperature of the water in tank, D, will fall until the medium of one hundred and two degrees is reached, when the alcohol in tubes, K, will contract far enough to lower the lever, O, valve, P, and regulator, I, to their former positions. X is a tube for drawing off the water from the cold water tank, B; Q is a slide, when the heat does not pass off as rapidly as necessary through the valve P, to cool the incubator space C, open the draw Q, and the cold air rushes in from below drawing out the hot air above. It should always be borne in mind, however, that the opposite end of the machine has a similar arrangement to that shown in figure 1, with which one of the tubes, K, connects—the whole operating in connection. The wires, R, are not rigidly attached to valve P, as above mentioned, but slide through a staple or orifice in the same, to the end that the lever, O, may have free play when the valve is closed or opened to its utmost extent. A similar result is obtained by the use of the spring, V, which permits the lever to rise after the regulator has closed over the tube to its utmost extent.

THE REGULATOR AND TUBE

are so arranged, however, that the flame cannot be entirely extinguished by the operation described, while the flame is graduated from a full blaze to a very faint one.

THE GREAT DIFFICULTY IN ARTIFICIAL HATCHING.

It is well known that the great difficulty in artificial hatching is that of maintaining a regular temperature, particularly in so variable a climate as in the Northern States. The difference of temperature between day and night has to be carefully provided for, and constant reference must be had to the thermometers. This difficulty has heretofore been a great obstacle in the way of the artificial hatching of chickens. By this invention, is obtained a constant and even temperature at all times, provided, of course, that the lamps are capable of producing sufficient heat for all exigencies.

THE PERIOD OF INCUBATION BY THESE MACHINES.

M. VALLEE, an inventor of one of these machines, in giving the result of his experience touching the period of incubation necessary for the various species of eggs, states what is curious and worthy of record. For chickens it takes twenty-one days; partridges, twenty-four; pheasants, twenty-five; Guinea hens, twenty-five; common ducks, twenty-eight; pea fowls, twenty-eight; Barbary ducks, thirty; geese, thirty; turkeys, twenty-eight. The degrees of heat required to effect the above result are from one hundred and four to one hundred and ten degrees Fahrenheit.

ARTIFICIAL HATCHING OF DUCKS IN CHINA.

In closing our remarks on the use of incubators and artificial mothers, we have deemed it not improper to give, from Commodore PERRY's report of his voyage to Japan, the mode used by the Chinese in hatching ducks' eggs by artificial means. After visiting the hatching chambers he carefully details the plan of the Chinese, as follows:—"There was

NO ARTIFICIAL HEAT IN ACTUAL USE

while I was there. The temperature of the external atmosphere was at about ninety degrees Fahrenheit, and there was a small chamber with a number of furnaces and charcoal, ready to be lighted and put into requisition at very short notice. The front room had large shelves on the two sides, about four feet deep from the wall, extending the whole length, the lower about a yard from the ground, and two others about eighteen inches apart. These shelves were appropriated to eggs which were within two or three days of their term. The shelves were first covered with two or three thicknesses of heavy, spongy paper, almost as thick as a blanket, which appeared to have been manufactured for the special purpose, in sheets four or five feet square. Next came a layer of eggs, two deep, all over the shelves, and two of the layers of the blanket paper mentioned. Parts of these shelves were occupied. They felt very warm to the hand. Their warmth was certainly much above that of the atmosphere, the blanket paper protecting them from its chilling influence as well as sudden changes. On some parts of the shelves the eggs were hatching, and the men were engaged, where they were nearly all hatched, in separating them. They tossed the little ones, as well as the eggs which showed signs of animation, very roughly and carelessly into baskets at considerable distance, greatly endangering the strangers' lives from concussion, fracture of limbs, &c., in our estimation, but in John's opinion it merely broke the shells, and thus enabled them the better to extricate themselves. The ducklings, after remaining a few hours to dry, and extricate themselves from the shells, were placed on the floor in little movable basket-work inclosures of bamboo, and supplied with a kind of grass chopped up for food, which they ate with an appetite which showed that they fully appreciated it. This grass was placed in little baskets with broad bottoms, so that they could not be overset, and the vertical splints continued upward, and were tied together at the top, so as to afford slats in the manner of a horse's manger. They could stick in their heads in the scramble for their first breakfast, but could not trample the food under their feet. I presume the young are transferred almost immediately to the boats, as I did not see any which appeared more than a week old.

"At the back part of their room is a mud wall partition, with a door in the center, and two other walls running back at right angles to it, dividing the back end of the building into three small apartments—one for the furnaces of charcoal, &c., the middle one serves as entrance, and the third is the apartment appropriated to the most delicate part of the process. This has a board floor, raised about four feet from the ground, beneath which are placed the furnaces, if necessary. The apartment itself was very dark and smothering; not much gas or smoke, but high temperature. This apartment contained about ten barrels, lined with the flannel paper, *stratum super stratum*, about three or four inches thick. In these barrels the process begins, and continues till within two or three days of its termination, when they go to the shelves in the front room. The barrels are almost filled with eggs, a sheet of paper being interposed between each layer of about six inches, and the whole covered with three or four sheets of the flannel paper, and a thick light lid, composed in part of the same material. The whole arrangement seems to be a most perfect protection from sudden changes of temperature, and I am under the impression that the eggs are handled a great deal, as they opened them without any hesitation, and even asked us if we should not like to invest capital in the business, for which they offered to pay two per cent. a month, or a share of the profits, which were certain to be equivalent."

From this description it appears that the first, and possibly the most delicate, stages of incubation are superintended with greatest care, and that the eggs are more freely exposed to the atmosphere as the incubation approaches completion. It is to be regretted that the exact temperature of the rooms is not given, the sensation of warmth being quite fallacious as a test of temperature.

REARING CHICKENS BY ARTIFICIAL MEANS.

ARTIFICIAL MOTHERS.

WHERE poultry breeding is carried on to a large extent, and where it is intended to rear the greatest number of chickens with the least number of hens, or with an incubator, artificial mothers are of the utmost importance. Chickens can be just as well reared, and, some writers aver, even better by artificial than by the natural method. The only use of the hen is to prevent the natural heat of the chick's body from cooling—to break up the food, and protect them from danger. In fact, chickens do not really require an artificial hen. They only require a suitable covering for their bodies until full-fledged, to preserve the natural heat, so as to keep their bodies warm, the same as full-grown fowls.

TO GET EARLY CHICKENS.

The artificial mother is very convenient to persons raising poultry, either on a large or small scale, to get early chickens in January or February, when the weather will not permit them to run out, and to have fine, large fowls for exhibition in the fall months. For large poultry dealers a good, light house is required, with good ventilation, without a draught; a dry and well graveled floor; sunlight, and a small run, with a little fire, in very cold, damp, chilly and rainy days, to keep the atmosphere dry, is all that is needed to raise as fine chickens as may be desired. The artificial mother, however, is a great economizer of time and labor—saves the necessity of any coops, which would otherwise be needed. It protects the little chicks from the changeableness of the weather, and from the vermin that infests, more or less, all poultry yards. By this mode the chickens are also completely under control, and where they can be given all sorts of nourishing drinks and food, without fear or trouble of the mother hen.

ARTIFICIAL MOTHERS MAY BE USED WITHOUT THE INCUBATOR.

Hens, especially those of the large or Asiatic breeds, are apt, when in confinement, to kill their chickens by treading upon them, or in scratching, and occasionally some mothers pick their young to death or prevent them from coming near her to pick up food. In the use of the artificial mother this can be obviated, and each chicken allowed to get its equal amount of food with the others. If the incubator is not used for hatching eggs, as we have said before, the artificial mother will be found to be a very useful appendage to the poultry yard, in more ways than one. As soon as the chickens are hatched out, say in about twenty-four hours, take them from the hen and put them into the artificial mother. Then place the hen back into the pen, in readiness to perform, in a few days, her ordinary functions in laying.

PERSPECTIVE SECTION OF A PORTABLE ARTIFICIAL HEN.

The engraving herewith given is taken from *Geyelin's Poultry Breeding*, and described as follows:—*A* is a glass-covered frame three feet long, fifteen

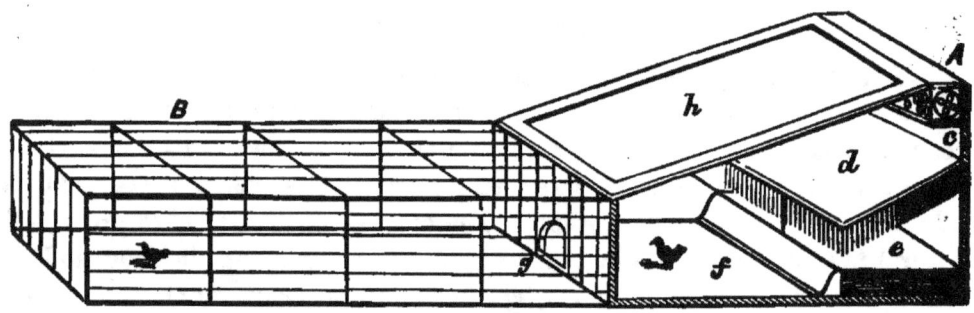

PERSPECTIVE SECTION OF A PORTABLE ARTIFICIAL HEN.

inches wide, two feet high at the apex, and twelve inches at the rise of the glass frame. This forms a dry run in wet and cold weather. *c* is an air-flue across the frame for the necessary ventilation, and formed of perforated zinc. At each end of this flue a ventilator is fixed, by which the admission of air can be regulated according to the temperature of the atmosphere. It will be apparent that chickens are not exposed to draught by this arrangement of ventilation. *d* is a frame lined with long fleece, under which the chickens will roost the same as under the wings of a hen, and will even prefer the artificial mother, as I have ascertained by experience. *e* is about one inch deep of ashes, which may be sprinkled over with flour of sulphur. They make a dry and warm footing, and retain the heat; but they should be renewed or sifted once a week. *f*, the floor, should be slightly covered with sand, and renewed every day. *g* is a small door, communicating with the open run. *h* is a glass frame, made to open by means of a slide or by hinges. *B* is the moveable open run, six feet long, fifteen inches wide, and twelve inches high. It is made of galvanized iron wire, which not only keeps the chickens from danger, but also prevents them from roaming. The artificial mother, being portable, should be taken in-doors every afternoon during the cold weather, and in the daytime should be placed on grass or dry land. The run should be made of small mesh, rat-proof wire.

GRAVES' ARTIFICIAL MOTHER.

This engraving represents one of the most approved artificial mothers of the present day, in fact, we believe, the best yet in use or invented. It is a box about six or eight feet long, with a glass door or lid that lifts up. The apparatus for heating the artificial mother is on the same principle as that of

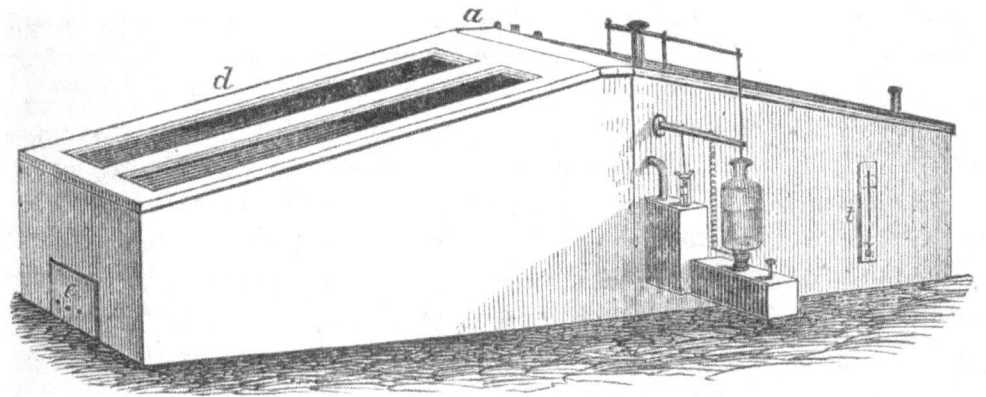

GRAVES' ARTIFICIAL MOTHER.

heating the incubator. It consists of a tank, *a*, filled with warm water, inclosed in the box under, *a*, and provided on its under side with a lining of sheepskin, or other soft material, and having an open space, covered with a glass roof, *d*. *t*, thermometer, regulating the heat on the inside. *e*, sliding door, for the chickens to run in or out, at either end of the artificial mother. As we have said elsewhere, the mother is heated on the same principle as the Graves Incubator, therefore we deem any further description unnecessary and superfluous.

IRREGULAR SEXUAL VARIATIONS OF PLUMAGE.

THIS is a subject that has of late attracted considerable attention of breeders, not only in this country, but also in England and France. Some time during 1869 a correspondent and particular friend of *Moore's Rural New-Yorker* entered a complaint against a well-known breeder, of having been swindled by him. He (the said correspondent) having purchased a pair of fowls and represented that instead of receiving a male and female bird he had got two male birds. Some time after he discovered that one of the birds had every appearance of being a cock bird, both in plumage and action, but laid an egg every day with the regularity of clockwork. He could not account for this curious freak of nature. He set the eggs of this product, but never had any of them hatch; seemingly none of them were impregnated.

We now get, through the London *Field*, illustrations of birds of this character, which we give in this connection, with a condensed report of the same as made to the *Field*, by the well-known author W. B. TEGETMEIER. He says:—"The case of assumption of male plumage by the female as represented in fig. 1, is certainly one of the most extraordinary on record, for the hen has not merely taken on the appearance of the male of her own variety, but has become still more masculine. Every poultry fancier knows that a Sebright bantam cock is what is called a hen-feathered bird—viz.: It has a square tail like that of a hen, and is destitute of the flowing sickle and saddle

FIG. 1—BARREN FULL-FEATHERED SEBRIGHT BANTAM HEN.

feathers and long pointed hackles that ordinarily distinguish the male species of domestic poultry. It might have been anticipated that the barren Sebright hen would have only assumed the male characters proper to the breed to which she belonged, such as a largely developed comb, elongated spurs, and the masculine crow; but this specimen acquired the long sickle and saddle feathers and pointed hackles of an ordinary full-feathered cock, still retaining the beautiful lacings or markings peculiar to the variety to which she belonged. The hen died in the autumn of 1869, before she had quite got through her molt, consequently the sickle feathers are not so long and curved as they would have been had she lived a few weeks longer. She was, however, carefully preserved for me by Mr. E. WARD of London, and the engraving is a very faithful representation of her appearance. The converse of the assumption of male plumage by the hen is the putting on the female plumage by the cock. There are, as is well known, several varieties of domestic poultry in which the cocks are hen-feathered, as in some breeds of Hamburgs and Game. This peculiarity is generally hereditary, and in the old days of the cock-pit, hen-cocks were well known. There is, however, a re-

FIG. 2.—FERTILE HEN-FEATHERED GAME BANTAM COCK.

markable distinction between the two cases described. A hen that has assumed the male plumage does so from being barren, and in consequence of disease or degeneration of the ovary. A hen-feathered cock, on the contrary, is perfectly fertile, and usually produces chickens with plumage like his own. The change of plumage from the full feather of the cock to the sober attire of the hen has never, I believe, been recorded, except by myself. It was a Game Bantam that was kept by me as a stock bird for his first season, and that changed at the second autumnal molt into the plumage of a hen of the same variety—namely, brown-breasted red. During his second breeding season, and as long as he lived afterwards, he produced chickens, some of which were full-feathered cocks, and some hen-feathered like himself."

POULTRY ENEMIES.

As every poultry-yard is more or less infested with, or annoyed by rats, weasels, skunks, and other vermin, we have been induced to give in these pages what we can find upon the subject of interest, and

HOW TO PREVENT THE DEPREDATIONS

of vermin on poultry. The most common enemy has proved with us to be rats. We have had them frequently carry off chicks and ducks fully a quarter grown, to say nothing of the depredations they have committed on broods but a few days old, in some instances carrying off whole clutches in the course of a few nights.

THE COMMON STEEL TRAP.

We have used the common steel trap, for catching rats, with good success in our poultry-yard, but after a time the varmints become shy of its open jaws, and it fails to perform the good offices we desire to have it. We have then taken to the

COMMON BOX TRAP,

which is shown in fig. 1, and with which we have been quite successful. It can be made by almost any one who is conversant with the use of tools, requiring a few boards, nails and wire in its construction, and will last for years, with any ordinary care.

HOW THE TRAP IS MADE.

The top and bottom of the trap are made of oak boards one inch thick and twenty inches square. It is divided into two parts, making really two distinct traps. The corners are of wire about one-quarter inch diameter, and the sides and partition of No. 7 wire. Holes are bored both top and bottom and the wires inserted. The corner wires are riveted, holding the trap firmly together; the doors are of oak, three-quarter inch thick, and are kept in place by a cross wire on the top board of the trap and by two small staples near the bottom edge of the door, which slide on the upright wires on each side. The treadle, X, is also oak, working on the upright pin, O, as a fulcrum, and being held in place by the wire hook, V, working on a pivot at

P, and on the lower end of which the bait is placed. One side of the trap is represented as set, the other as sprung.

SETTING AND BAITING THE TRAP.

In setting this trap, when the rats are abundant, we have always baited the trap for several nights before setting it in earnest; we fasten the bait to the hook, and then fix the trap so it cannot be sprung, then strew Indian meal or other feed around the bottom of the trap. In a few nights the rats will make this quite a feeding ground. We have caught, says a writer in *Moore's Rural New-Yorker*, twenty-seven rats in a single night; sixteen at the first setting and eleven at the next. Then perhaps it would be a week before we would catch another in that trap; meanwhile we would start another.

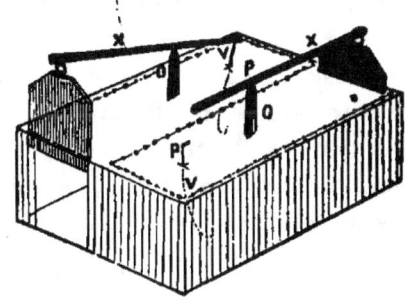

Fig. 1—Common Box Trap.

THE BOX OR BARREL TRAP.

One simple arrangement has caught scores for us. In any building or cellar where the rats abound put a water-tight box or barrel; if a box, it should not be less than two and a half feet deep; about one-third down from the top hang a lid or trap-door, hanging it from the side of the box or barrel. (See fig. 2.) Cover this lid with a piece of tin or sheet iron in such a way that there is no roughness to make a foothold for rats. To hold up this lid, make a common wire spring, thus X, X, passing through the side of the box or barrel, to the ring of which attach a cord; carry this cord to the outside of the building or cellar, so that it can be pulled without being obliged to enter the room where the trap is. The lid should hang so as to drop, not lift or raise. Sprinkle some corn meal or other feed on the lid, having previously put about six or eight inches of water in the box. At any time during the day or night, when you are passing, pull the spring and drop the lid; a minute's time will reset the trap, and, although you may often catch nothing, you will sometimes catch half a dozen at a time. We have known over a dozen caught during a single evening, and in the course of a month a house almost depopulated of rats.

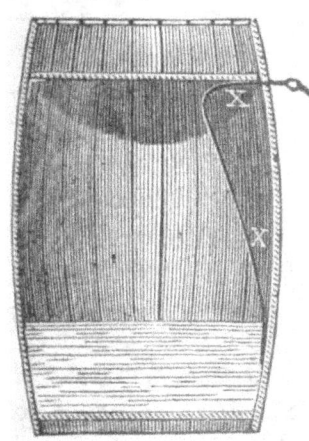

Fig. 2—Barrel Trap.

WEASELS, MINKS AND SKUNKS.

In the country we also have weasels, minks and skunks to fight against. If the place abounds with mice the weasel will rarely touch the chickens, the former being his favorite food. But when the weasel once gets a taste of chickens, he will sometimes slaughter whole broods in a single night; he simply sucks the blood and passes on to the next. We have known them to attack full grown fowls, but rarely; unless their burrow is near by, they will seldom visit the same yard two nights in succession.

THE RAVAGES OF THE MINK.

Next to the weasel, the mink is most dreaded among poultry. In localities near salt marshes, swamps, ponds and sluggish streams they most abound. The ravages of the mink are easily told from those of the weasel, or any other animal. He almost always carries off a portion of his prey and tries to secrete it. If you find a half-grown chicken or old fowl dead and dragged wholly or partly into a stone wall or under some building, you may be certain it is the work of a mink; and if you go to work right, you will be just as certain to trap him.

THE PECULIARITIES OF THE MINK.

One peculiarity of the animal makes his capture easy—he always returns to a spot where he has hidden his quarry, or where he has made a raid; and if he misses it, will go searching around for it. A knowledge of this fact led to the invention, some ten years since, of the trap given in fig. 3. The trap should be three feet long, one foot wide, and one foot high, outside measurement, and may be made of ordinary faced pine boards. N is the only solid part of the top, to which is hinged the lids L and D, and also in which the standard S is mortised. The lid L is held up by the rod A, in which are one or more notches, to elevate it

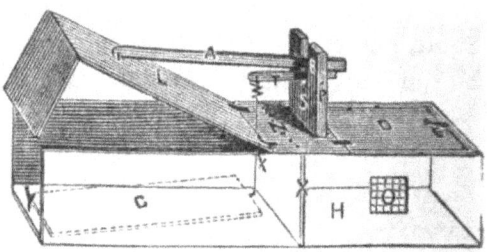

Fig. 3—MINK TRAP.

the desired hight, catching or hooking over the pin B, and projecting a few inches beyond. Under A, and hinged into the standard by the pin P, is the lever T, also projecting an inch or more beyond. C is a treadle-board, hinged at Y to the bottom of the trap, and connecting by the wire W to the lever T, elevating it about two inches when set. H is the bait box, separated from the main trap by a wire screen, X, X. O is a window, of which there should be one on each side about three or four inches square, also covered with wire or wire cloth, and D is the lid of the bait box, fastened down by the pin E.

BAITING AND SETTING THE TRAP.

If you have a chicken or fowl that has been killed by the mink a night or two preceding, put *that* into the bait box and close the lid, placing the trap as near the spot where the dead fowl was found as you can. If a live fowl is put in, no harm can be done to it, the screen effectually protecting it. The mink enters the trap, and as soon as his weight gets well up on the treadle it pulls down the lever T, the projecting end of which dislodges the rod A, and drops the lid L. It is best to have a *weight* upon L, or else a catch to hold it down when sprung, as we have known an old mink to pry up the lid and get out. We have never known this trap to *miss* when set immediately succeeding the depredations of one of these *varmints*.

YOUNG MINKS SUCK EGGS.

Young minks not one-third grown will suck eggs. A friend of ours once found three young ones in his stable, each with its head inside of an egg shell, and as effectually trapped as any one could wish, which he soon dispatched.

DESTRUCTIVENESS OF THE SKUNK.

Next to the mink, the skunk is the most destructive to poultry. We have had three entire broods, thirty-seven chicks with two or three hens, killed in a single night by these animals. We at that time, some fifteen or more years ago, put our hens and chickens on the bare ground. The skunk dug under, and then had the fun all to himself; since then we have made all our coops with hard bottoms, and have lost no more chicks from that cause.

HOW TO BAIT THE TRAP FOR SKUNKS.

The only way we have trapped the skunk was with eggs, of which they are passionately fond. Neither are they particular about the quality, as they seem to favor a rotten one, or one with a dead chicken in it, as well as the best and freshest. Tie the egg in a piece of netting, and fasten it to the

Fig. 4—Barrel Skunk Trap.

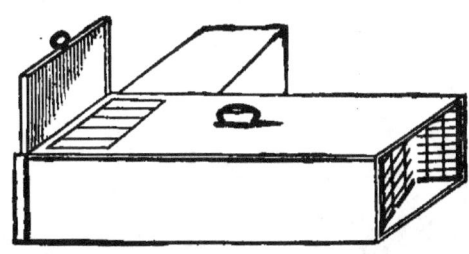

Fig. 5.

treadle of a steel trap, or to a common box trap. Take care that it is a trap you do not wish to use again *soon*, for it will doubtless be too highly flavored to suit a refined taste.

MANNER OF SETTING THE TRAP.

Find their burrow and set your trap near the mouth. It is nearly useless to set a trap where a theft has been committed. The animal may not go back there again for months. He might possibly be caught in a night or two, but the chances are ten to one against it.

MR. HUNGERFORD'S SKUNK TRAP.

A correspondent of *Moore's Rural New-Yorker*, from Lyme, O., says he succeeds in ridding himself of skunks in the following manner, which certainly is a simple mode. He says:—"I take an old barrel (see figure 4,) and place it on a stick of wood about seven inches high, as shown in the sketch. When the skunk goes for his bait, which is in the bottom of the barrel, as soon as he passes the center the barrel turns up with the skunk, woodchuck or other animal caged in the latter, without making any scent. I then take Mr. Skunk by the tail, and carry him where I please. Care must be taken not to place the barrel too high, as it might throw over and not remain upright. The more skunks you catch in the same barrel, the better the trap."

GEYELIN'S VERMIN TRAP.

"The trap (see figure 5) consists of an oblong box, the end of which draws out, and is provided with a looking-glass in the internal side, which attracts the vermin on looking in. The entrance of the trap is formed of two spring doors made of wire, which allow the vermin to enter with the least pressure. These doors have sharp points where they meet, which, although not felt by the vermin on entering, will prevent it from withdrawing after having once introduced its head. Near to the looking-glass a bait is suspended, and a cage is also fixed with a chicken to serve as a decoy. These traps are self-setting, simple, inexpensive, fit for all sizes of vermin, and safe for the house, farm-yard, or game preserve."

H. MILES' VERMIN TRAP.

Mr. MILES says, the best trap for farmers to catch rats, mink, weasels and skunks ever set is:—"Take boards half an inch thick, and make a box the two sides and top twelve inches long, with one end closed; the size of the box inside being four inches square. I give you a rough sketch of the

FIG. 6.—THE TRAP SET.   FIG. 7.—THE TRAP SPRUNG.

trap as set. A steel spring is fastened on the closed end of the box, to which is fastened a square ring at its extremity, through which the game thrusts its head to reach the bait at one end of a catch, which holds the ring depressed, and held by a wire running from the front end of the trap to the catch on the upper extremity of the bait hook. This is the best trap for skunks in the world, I believe. I have used many different kinds, but none works so well as this. You can set it at a hole in a wall or fence. It is sure fire."

CROWS AND HAWKS.

Crows and hawks are to be classed among the enemies of poultry. The former prey only on young chickens and eggs. Catch one and hang it in your poultry yard; no other crow will come near it.

HOW TO TRAP CROWS.

The quickest and surest trap for crows is to place a steel trap in the shallow water of a pond, so that the jaws, when open, are just under water. On the treadle place a small tuft of grass or moss, making a miniature island. Then cut a small stick with three branches, forking in such a manner as to support an egg on them; stick this about six or eight inches from the trap; lay a little moss, grass, or leaves over it, and place the egg on the forks, so it will appear as if floating on the water; cover the remainder of the trap lightly with grass, so as to hide it from sight. To obtain the egg the crow will light on the "*island,*" and find, too late, he's caught.

HOW TO GET RID OF HAWKS.

When hawks are troublesome, the only remedy is to shoot them. You will soon notice that he visits your yard about a certain time every day, and by watching for him you can soon rid yourself of the troublesome visitor — of course, provided you are a good shot.

PACKING EGGS FOR TRANSPORTATION, ETC.

PUTTING THE LARGE END DOWN.

A new fact has just been developed in regard to the packing of eggs for transportation and for hatching. The old theory of packing eggs with the small end down has been practiced so long that many think that the infallible mode. But the experiment we are about to relate confirms us in the belief that the *modus operandi* now recommended is a good one, and, coming from the source it does, is worthy of consideration by those desiring eggs transported to them from a distance. The article in question is from the pen of L. WRIGHT, author of the *Practical Poultry Keeper*, and those familiar with his writings on poultry or conversant with the reputation of his book, need no other assurance of the feasibility of the results arrived at by him. He says he has discarded bran in packing eggs, and substituted therefor hay; believing that good, soft hay is the best material that can be used.

MANNER OF PACKING.

His mode is to put a good layer of hay in the bottom of a box, not rammed down, but left springy, and some hay put all around the inside; the eggs should be nicely bedded in one layer only. Each egg should be wrapped singly and loosely in a piece of paper a quarter the size of a common newspaper page, in such a manner as to leave the ends square, and not shaped to the egg. Then a good wisp of hay is wrapped round each, and the eggs put in the box just tightly enough to prevent them from shaking about, and no more. The eggs should be packed with the *large end down*. Mr. WRIGHT says he can state *positively* that eggs intended for hatching will keep good much longer and better when placed on the large end than in any other position. Mr. GEYELIN also advocates this position for eggs intended to hatch.

THE REASONS FOR PACKING WITH THE LARGE END DOWN.

Mr. WRIGHT's experiments in this matter extend over a period of two years. A lady correspondent of his, of large experience, writing him upon this subject, says:—"Keeping eggs on the small end appears to me to cause the air-bubble to spread, detaching it from the shell, or rather from its membraneous lining; and after being so kept for a fortnight the air-bubble will be found to be much spread, and the egg to have lost much of its vitality, though still very good for eating." In describing her success with keeping eggs in a contrary position, (large end downward,) says:—"Owing to this method of storing, such a thing as a stale egg has never been known in my house; and as regards success in hatching, for several seasons when I was able to attend to my poultry myself, of many broods set, *every egg* produced a chicken."

CORRECTNESS OF THIS THEORY.

Again, to prove that Mr. W. is correct in his theory, he cites a case wherein he shipped thirty eggs from England to a gentleman in Ohio, packed in the manner here described; the eggs were twenty-two days on the way, and eighteen chickens were hatched from them. In regard to these eggs he says:—"As I had not many hens laying at the time, many of the eggs must have been eight to ten days old when sent, and fully a month old when set; and I think, therefore, the simple fact that they hatched in the proportion of six to every ten will be sufficient warrant for my now recommending to other fanciers, with full confidence, the adoption of this position for packing and storing."

Mr. BABCOCK, New-Haven, Conn., sent us twelve Muscovy duck eggs packed in nearly the same manner as described, which were set under a barn-yard hen; the eggs becoming chilled they did not hatch, but on being broken we found eight of the twelve had dead ducks nearly fully matured in them. We have, therefore, no hesitancy in recommending to those desiring eggs for incubation, to request that they be packed in the manner here described, and in no other way, as we believe more than one-half the complaints that arise about the unfertility of eggs, can be traced to no other cause than the careless, bungling and improper manner in which they are packed and shipped.

A WORD ABOUT PACKING BOXES.

We have used and seen a great many packing boxes, but think the one we here describe, (and which is not patented,) is one of the very best, and which received the premium at the annual exhibition of the New-York State Poultry Society. This box is intended for the transportation of eggs of fancy fowls desired for hatching purposes. It is described as follows: It is a box made of pine wood, dove-tailed together, ten and three-eighth inches long, eight and two-eighth inches wide, and six inches in hight, containing twelve compartments of wood, which are fastened together so that they can be pulled all out at once. These compartments are two inches wide, and nearly the depth of the box. To pack the eggs you pull out the inner boxes, and place bran, cut hay or any other soft substance in the bottom, and then replace your box compartments; then roll your eggs in soft white tissue or tea-paper, wide enough so that when you turn down the ends they will lap over each other, and so doubly protect the ends of the eggs, and then wrap them in newspaper, so as to make almost a small square bundle, folding the ends over nicely. Then place this, with the large end of the egg downwards, in the several compartments, until they are all filled. Then place the bran in around the edges or open spaces, so that the egg cannot move, and fill the top up with bran. Place on the cover, which slides into a groove on either side of the box, and screw the end down with one screw, so that the cover will not slide off. Then place your box in as many newspapers as you may think proper for the distance it is to go, and cover the whole with a covering of thick hardware paper, and tie with a strong thick twine, for a handle.

MODE OF PACKING EGGS FOR MARKET.

As we have given a description and manner of making boxes for the transportation of eggs, for hatching purposes, we now propose to give a description and engravings of two new and useful inventions, by which eggs may be carried any required distance with perfect safety, at all seasons of the year, without loss from breakage or heating. By this arrangement the present system of packing eggs in barrels, or other packages, by the use of oats, chaff or other material is entirely abandoned.

THE CANVASS-COVERED CASE.

Figure 1 represents a substantial carrying case, with nine draws, the frames of which are of wood, covered with canvass or sacking, with cords or strings underneath, for the purpose of keeping the eggs in their places. The sacks, at the top and bottom, have depressions, as shown in the cover of the engraving, so that the eggs fit snugly, and are not liable to be displaced by handling or transportation. Each alternate layer, coming between these depressions in each box or drawer, fills up the interstices perfectly. With proper care these cases will last for years; are always ready for packing, and can be filled as the eggs are laid—thus avoiding repeated handlings, which so frequently injure them. The eggs can also be kept in them perfectly secure, when the owner desires to hold his stock for a better market. There are nine layers or drawers of eggs in this box, each layer containing eight dozen, or a total of seventy-two dozen of eggs.

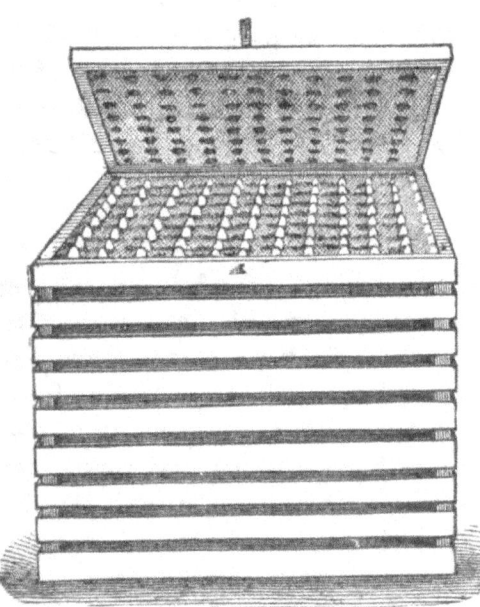

Fig. 1.—Canvass Covered Case.

THE COMMON TRANSPORTATION CASE.

Figure 2 shows a cheaper case, in every respect. It is a common packing box, made with paste or binder's-board partitions, and each layer of eggs is

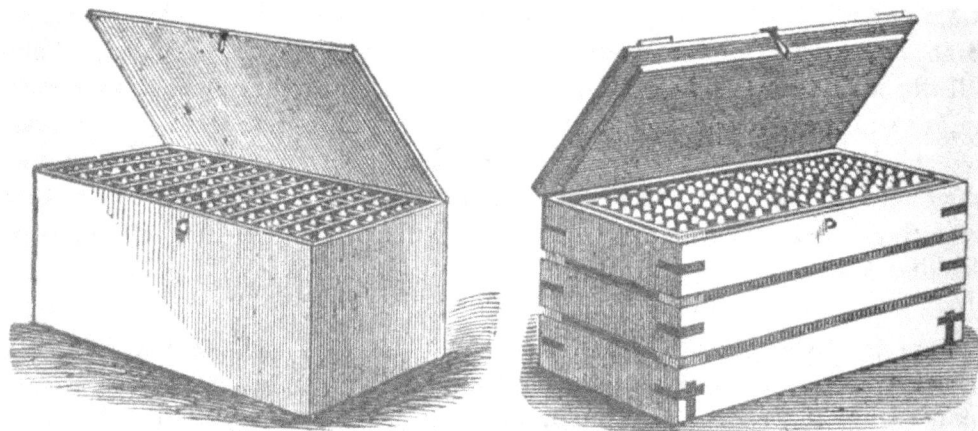

Fig. 2.—Common Transportation Case.   Fig. 3.—Suspension Egg Carrier.

covered with the same material. One point connected with packing in these boxes the shipper should know and guard against, that is, it is sometimes the case that the paste-board cover, on which the eggs are placed, is composed of two pieces, and during transportation or handling these pieces become displaced, or pass each other; then the eggs above drop down on the lower ones, and break them. This difficulty, however, can easily be avoided by pasting a piece of stiff paper over the joint, which will prevent them passing each other. Should there be too much space in the top of the case, place a paper or two over the eggs, and fill with straw or hay, which will keep all tight and compact, without any fear of breakage. Any sized box desired can be used for this style of case, and, with a little care on the part of the packer of the eggs, can be carried as safely as with any of the patent boxes or egg-packing cases now in vogue.

SUSPENSION EGG CARRIER.

Figure 3 consists of an outside case or crate, in which are fitted a number of trays, with cords laced through the sides and ends, dividing the space into small squares or meshes, and making a delicate spring, which responds to the slightest jar. Rows of pockets are suspended from the cord work, giving to each a separate apartment, and so arranged that no jar nor jolt the carrier may receive can cause one egg to strike another; and being thus seperated, a free circulation of air is obtained, which prevents heating by any possibility. Each tray is provided with a protector, which keeps the eggs in the pockets even though the carrier be overturned. As each tray contains a certain number, no errors in count can ever occur, and the purchaser can determine at a glance both the number and quality of the eggs. By using the carrier a child can pack as well as a man, and much faster than by the present system. One of these carriers, the size shown in the illustration, will hold sixty dozen eggs.

THE CHAMPION EGG CARRIER.

This carrier consists of a case or box, with a tight cover, and trays fitted inside, each to contain three dozen eggs. The trays are made of strips of straw board, folded double, and so arranged as to make pyramid pockets, smaller at the bottom than the top. By an ingenious invention these pockets are provided with delicate springs in each corner, at half the depth of the pocket, and where they are not subject to wear, which causes the bottoms of the double strips of paper to spring apart or diverge from each other, so that eggs being placed in the pockets on their small ends, although of various sizes, are held in an upright position, and no jolt or jar can cause one egg to strike the other.

PRESERVING EGGS.

We have tried several modes of preserving eggs, and never found any difficulty in keeping them any length of time, for culinary purposes, nearly as good and fresh as when first laid. Our principal mode has been that recommended by Mons. Chas. Jacque, which, from several years experience, proved one of the best we ever tried; having succeeded in keeping eggs nice for use from six to eight months after they were laid.

THE MODE OF PUTTING DOWN.

The most certain and most lasting mode of preservation consists in covering the eggs in a jar filled with lime water, recently prepared, and keeping them in a cool place. The lime water is prepared from quick lime, or that which has been slaked but lately, by placing it in a quantity of water greater than would cover the eggs. The milk of lime which is thus formed is allowed to stand several hours. The clear liquid which separates itself from the excess of lime used is the lime water, which is poured off for use. Lime water not only prevents the evaporation, since the eggs are plunged in the liquid, but the alkali which it holds in solution closes the pores of the shell and prevents all fermentation, either of the eggs or of the organic matter which the water might contain.

PACKING IN SALT.

We have had good results also from packing eggs in very dry barrel salt, which have kept for months in a well preserved state. Our mode was to stand a box or stone jar in a cool place in the cellar, put therein a layer of salt, then one of eggs, with the large ends downward, taking care that the eggs did not touch one another; continue this practice until the jar or box is full; cover the box and let it stand without disturbing until the eggs are needed for use. We have packed eggs in this way in June and July, and found them in January and February perfectly fresh in looks and having no stale or musty taste when brought to the table.

THE FRENCH MODE OF PRESERVING EGGS.

The French mode of preserving eggs is to dissolve four ounces of beeswax in eight ounces of warm olive oil; in this put the tip of the finger and anoint the egg all around. The oil will immediately be absorbed by the shell and the pores filled up by the wax. If kept in a cool place, the eggs, after two years, will be as good as if fresh laid.

EGGS AS A COMMERCIAL COMMODITY.

THE CONSUMPTION OF EGGS IN NEW-YORK CITY.

Eggs form a very important commercial commodity. What the extent of the business is in New-York there are no returns to enable one to state definitely, but that it is large is manifest. We can form some estimate of it by glancing at the consumption in other cities.

EGGS CONSUMED IN PARIS.

In Paris it was calculated that the annual average consumption of eggs per head was one hundred and seventy-five, or in a population of two millions a total of three hundred and fifty millions. The average of the country districts was placed even higher than this, while the aggregate French product has been estimated at between seven and eight thousand millions annually, a number large enough to form a string of beads that would twice encircle the globe.

IN GREAT BRITAIN THE QUANTITY CONSUMED

has been set down at one billion five hundred millions, of which one hundred millions are imported. The bulk of the importations and large quantities of the home produce go to London, which, however, does not eat, proportionately, so large a number as Paris.

COMPARISON WITH NEW-YORK CITY.

It will be safe to assert, after a due comparison with these cities, that New-York consumes annually one hundred million eggs — an amount which may be more readily comprehended by remembering that this number would weigh about thirty-six thousand tons, and on the principle that an egg is equal in nutritious value to a quarter of a pound of meat, would represent an average of twenty-five pounds of flesh meat per year, for every man, woman and child in the metropolis.

THE CASH VALUE OF THIS PRODUCT.

The cash value of this product cannot be less than two millions of dollars, while the demand for the article and its selling price are steadily increasing.

These figures show that trifling in detail as the egg trade may seem, in the aggregate it is an important business. And they suggest, also, the question,

HOW CAN THE SUPPLY BE INCREASED?

Unlike a great many other occupations, no one seems to make egg producing a specialty, with a single eye to making a living out of it; and yet, taking the figures above quoted as trustworthy data, they point to the conclusion that there's money in the business. Some years ago we read an interesting account of an establishment near Paris for the manufacture — or perhaps production would be a better word — of eggs and chickens for the Parisian market. It was on an immense scale, and was a great success. Why cannot we have large henneries near each of our principal cities? Or, to limit the inquiry to a single case, why cannot some enterprising person or persons, for there is room for dozens, establish one or more egg manufactories on the outskirts of New-York? Let us see what preliminaries would be necessary for such an undertaking.

AMOUNT OF CAPITAL TO BE INVESTED.

A fair but not excessive amount of capital, which would be invested under these two favorable circumstances, that there would be immediate returns for the outlay, and an excellent market for the produce.

A SUPPLY OF HENS.

But it is not necessary that these should be of any fancy and, consequently, high priced breed, another important consideration affecting the original capital. General experience has shown that good barn-door fowls, as they are termed, prove as profitable in the end as more pretentious birds.

A PIECE OF LAND PROPORTIONATE TO THE STOCK OF HENS.

Mr. WARREN LELAND of the Metropolitan Hotel, New-York, finds it beneficial to allow an acre to every hundred hens, but rough, broken ground, with some low bushes and heaps of sand, ashes and lime scattered about, answers admirably. The London *Field* corroborates Mr. LELAND's experience, while a writer in the *Massachusetts Ploughman* states that six acres is plenty for a thousand hens. If the land is to serve in part as a feeding ground, then the first estimate; if merely, or mostly, for exercise, the latter is ample.

A GOOD SUPPLY OF FOOD.

This is important. "Hen-laying," says Mr. LELAND again, "is hard work, and requires high feeding." This is very true, and a false economy here would be fatal. But he points out also a cheap and excellent source of supply. "Much of my success," he adds, "is due to the fact that my hens get all the scraps from my hotel." In a large city, like New-York, these scraps could be obtained for a trifle in many cases, for the mere trouble of collecting them in others. Why cannot they be gathered and transformed into eggs? This will prove the corner stone of success in making eggs cheaply. Food of any kind always goes farther when cooked, and animal food and a certain degree of warmth are essential to early and prolific laying.

OTHER DETAILS WILL READILY SUGGEST THEMSELVES.

There must be sheds for the accommodation of the poultry without undue crowding, and these must be kept perfectly well ventilated and artificially heated in winter. Coal is a cheaper fuel than any kind of food. Layers, sitters and fattening birds must be kept separated. Quietness and cleanliness should reign everywhere, and all outside intruders must be rigorously excluded. Plenty of fresh, pure water is a necessity, and a portion of the old stock should be regularly renewed, as only from young and vigorous fowls could the best results be obtained.

CHICKENS NOT BROUGHT INTO THE ACCOUNT.

We say nothing of chickens, though these would doubtless form a useful department in the outline here faintly limned. Nor do we enter into any calculations as to the amount of profit to be realized. This would depend largely on personal qualities. A recent writer in one of our farming papers claimed a profit of ten dollars per hen per annum on a small scale. We believe that with New-York to furnish the scraps for almost nothing, and buy them back in the shape of eggs at a good market price, a greater average may be achieved; and are certain that a much lower figure would be largely profitable.

CARE OF POULTRY IN WINTER.

The Farmers' Club of the American Institute appointed a Committee to visit WARREN LELAND's farm, and examine his mode of keeping poultry in winter. The following is the Committee's report, made in February, 1871:

We spent a day at the farm of WARREN LELAND, twenty-five miles north of New York City, at Rye Station, and have derived, from a careful survey of his yards, ideas which we consider important. We find him carrying one hundred and fifty turkeys, about three hundred hens, a large drove of ducks, and several dozen of geese through the winter, without the loss of any of his poultry by disease of any sort, and without the freezing of their feet or of their eggs. We learn that he never has maladies among his poultry; that he will allow the greater part of his hens to sit in the spring, and each of them will yield an average brood of ten chicks, so that he will raise about three thousand chickens from his present flock, and his losses be very few. How does he do it? His hens, ducks and geese have the best winter quarters we have ever seen provided for any of the feathered tribes. Their

MAIN BARRACK, OR HENNERY,

is a stone house, seventy-five feet long, and twenty feet wide, and faces south. The openings on the north side are small, and filled with window glass, and in some cases with double sash. Those on the south side are much larger, consisting of double doors, which are opened on sunny days. In the middle of the north side is a wide, old-fashioned fire-place, with crane and a big camp kettle. Nearly every day in winter a fire is lit, and fed with chunks, knots and old logs, that would otherwise be knocked about the wood-yard, and left to rot in fence corners. The walls are of stone, and the floor of rock or earth, so the fire can be left without the least danger.

On cold days, and especially in cold rains, the hens gather before this fire and warm themselves, and trim their feathers. The chimney can easily be closed, or the logs rolled out into the middle of the building, and feathers or sulphur used to make a

FUMIGATION.

This is done whenever hen lice appear; and the openings of the house can be closed, so as to hold the fumigation till it penetrates to every crack. Smoke he finds better than carbolic acid or kerosene, or whitewash, to drive vermin. The roosts are oak slats, an inch thick by two and a half inches wide, fastened to the rafters near the ridge. They are nailed at different hights, and at proper intervals. About two feet below the perches is a scaffold of boards, that fit quite closely. This is from time to time covered with plaster and ashes. About once a month the accumulations are shoveled down, and piled up for the corn-field. He calculates that fifty hens yield, in the course of a year, as much compost as would be worth fifty dollars in bone meal; that is to say, if he threw away his hen droppings and had to buy the same amount of fertilizing salts in bone-dust, it would cost him fifty dollars to replace fifty hens as producers of manure. He has paid special attention to the comfort of his

HENS ON THE PERCH.

They sit on a slat two and one-half inches wide. Their breast feathers come down and cover their feet, and protect them from freezing in the coldest nights. Of course, there is no lack of dry ashes in their house, and he finds that after the fire goes out the hens use the hearth as a place to nestle, and shake ashes through their feathers. They enjoy it, and it keeps them sound and comfortable. The offal of the farm, as entrails, feathers, heads, scraps from lard, and all the odds and ends from the kitchen are thrown into this house, and the hens pick it over, eating all they want. Then, as soon as spring opens, all this trash is shoveled and scraped out, composted and taken to the corn-field. Besides this refuse, his poultry eat about a bushel of corn a day in winter, and half a bushel in summer. He raises large crops of corn, because he has strong manure to feed his crops with, his calculation being that about four acres of corn go to feed and fatten his poultry. In spring,

AFTER A HEN HAS HATCHED,

her nest is taken out, the straw burned, and the box whitewashed inside and out, then filled with fresh straw, and put back for another family party. After many trials of breeds, he has settled upon the white Brahmas. They lay more uniformly the year through; make the best mothers, and the chicks grow the fastest. During summer his poultry have a wide range, and scour the fields for half a mile or more,

CONSUMING GRASSHOPPERS,

His turkeys nearly make their weight on grasshoppers and beetles, with a handful of corn night and morning. One man has little to do in spring and summer but to take care of chickens and young turkeys. In winter they require but little attention, and this man then attends to the calves and lambs.

THE COST OF HIS POULTRY MEAT,

and he often kills in a season three hundred turkeys and three thousand chickens, he considers to be about two hundred and fifty bushels of corn, and the wages of his hen-wife for half the time. His gains he cannot give exactly, for the poultry is eaten very freely by a large family, and sent to the Metropolitan when prices are high, or the supply in market defective in quality. He does not keep exact account of his eggs, for as a rule he says the best thing to do with an egg is to let a good motherly hen make a chicken of it. Your Committee conclude their report by an expression of opinion that

THE COMMON IDEAS ON THE SUBJECT OF POULTRY RAISING,

on a large scale, are erroneous. It has been said, again and again in this Club, and in farm journals, that there is no use in trying to keep more than about fifty hens. If one goes deeper into the poultry business there is backset from lice, and roup, and gapes, and cholera, and the sudden death of hens and chicks from causes unknown. This is a fallacy. In the manner above described, by the wise use of smoke and lime, and ashes, and a fire, by cleanliness and a wide range in mild weather, we find Mr. LELAND taking about four thousand feathered animals through the season, for year after year, without calamity or loss, and on an expense that is very trifling, and unfelt on a large farm. Your Committee will visit other farms, where the special object is eggs, and announce the result of their observations. Dr. SMITH, a member of the Committee, said that, in conversation, Mr. LELAND said that his success depends upon letting his poultry alone. He repeated it over and over again, "let them alone; let them alone; give them liberty, and they will take care of themselves. Dr. TRIMBLE, another member, said that Mr. LELAND told him that in

RAISING TURKEYS

his plan was to have three or four sitting at the same time. When they had hatched the eggs, he gave all the young turkeys to one hen turkey, and she and her enlarged brood were removed to a distant part of the farm, away from other fowls. There a large coop was built, in which they could be shut up at night. They were not allowed to range in the morning until the dew was off. In the day time the hen was tethered to a stake; and each day her stake was removed to a new place, so that she and her brood had a new range.

A SOUTH AMERICAN POULTRY FARM.

Just as we were about closing the last form of this work for the press we received from C. F. PEARCE, Esq., Freetown, Mass., the following article giving an account of "A South American Poultry Farm," which we deem of sufficient importance to give in this connection, as containing valuable information upon a subject which will interest all breeders of poultry:

PROFITABLE POULTRY REARING ON A LARGE SCALE.

Although we have column after column of poultry matters offered for our perusal, through the agricultural press, I have as yet to see the first statement giving the facts and figures of profitable poultry raising on a large scale. Perhaps there might be some who have achieved success in this line, but we are led to believe that misfortunes are more plentiful than the fortunes acquired from the manipulations of this particular stock. I have tried my hand at the business, although on a limited scale, and can show figures giving me a profit of three hundred per cent. on the capital invested. There is no known reason why it cannot be managed in an extensive way, and with just as good results, provided it is entered into knowingly and understandingly.

DESCRIPTION OF A FARM WHERE POULTRY ARE KEPT BY THE THOUSAND.

I propose to describe a poultry farm, where fowls are kept by the thousand, and whose proprietor counts his gains therefrom proportionately. It is situated in the southern extremity of Chili, South America, where the rainy season, of six months duration, is as detrimental to the well being of all fowl kind as the rigors of our own winters, and where great care and skill is very essential to satisfactory results.

OPERATIONS COMMENCED WITH TWO HUNDRED HENS AND EIGHT COCKS.

Senor Don SAN FUENTES commenced his operations in poultry with a stock of two hundred hens and eight cocks, to which he has added thereto, by natural increase from year to year, until now he has somewhere in the vicinity of six thousand. Their range is unlimited, as his farm covers three thousand cuadras, equal to seven thousand five hundred acres. To every fifty hens and two cocks is given a house of their own, of which there are six or seven hundred on the place. These are placed two hundred feet apart, each way, thus isolating one lot from the other.

DESCRIPTION OF THE HOUSES.

These houses are very cheap affairs, and are made by erecting two forked posts, eight feet long, and distant from each other fifteen feet. On these rests the ridge-pole. On both sides of the centre-post, ten feet distant, a trench is dug, a foot in depth. Then small poles are placed for rafters, one end in the trench and the other tied to the ridge-pole, two feet apart. Then another set of poles, tied crossways, also two feet equi-distant, and the frame work is complete. This is covered over with thatch, which is found in plentiful abundance, and to be had for the cutting. The only frame work about the house is the doors at the ends, both of which are four by six, and contain each a window, pivoted in the centre of the sash, to be opened or shut as the requirements of ventilation demand. Each house has its complement of twenty boxes, for laying, placed under the eaves, and partly concealed by bundles of straw.

BUILDING FOR STORING GRAIN, EGGS, HATCHING AND SICK ROOMS, ETC.

Near the family residence is a large building, devoted to the storing of grain and eggs; a nursery for sick hens; a long room for hatching, and another for slaughtering purposes. In the sick room is arranged a series of boxes, each one large enough for the comfort and convenience of its solitary occupant, who is there placed, and treated for its malady with as much care as if its value was dollars instead of cents, and with such skill that the ratio of deaths has been only one in two hundred and eighty.

THE SITTING DEPARTMENT.

is also provided with boxes, some three hundred in number. Here all are brought, from their respective coops, as soon as their incubating propensity sows itself, and placed upon their quota of eggs. Feed, water, and a large supply of sand and ashes, are provided, and the sitting hen not allowed to leave the room until she takes her young brood with her.

HOW THE CLUTCHES ARE DOUBLED UP.

The clutches are then "doubled up," that is, two broods given to one hen, and the chickenless one sent back to her coop to resume her egg laying. As soon as the young chicks are discarded by their mother they are taken to their future home, fifty in each lot, and the old ones back to their respective localities.

HOW THE FOWLS ARE FED.

The fowls are fed three times per day, and their diet so arranged as to always present a variety, although oats is their staple article of food, and always before them in unlimited quantity. To-day, it will be indian-meal, made into a stiff dough, and given hot; to-morrow, barley; next day, boiled potatoes, mashed, and mixed with pork scraps and bran—corn broken in a coarse mill, and so on in rotation; adding from time to time a dead horse, or some other cheap and inexpensive animal food. Burned bones, pounded shells, and lime, are supplied in profusion, These, with what they gather on their foraging expeditions, produce a wonderful supply of eggs.

NOT ALLOWED TO LEAVE THEIR COOPS IN RAINY WEATHER.

During the rainy season they are not allowed to leave the coop, except the day be exceedingly pleasant, and then only for a short time. They appear to bear their confinement remarkably well, and with hardly any decrease in the quantity of eggs. While confined they are allowed an extra allowance of animal food.

ATTENDANCE REQUISITE TO THE CARE OF SIX THOUSAND FOWLS.

The attendants requisite to the care of these six thousand fowls are one man and four boys. The houses are thoroughly cleaned once a week, and the interiors whitewashed every three months. Every morning each lot of fowls undergoes a careful inspection, and any one found moping or otherwise indisposed is immediately taken to the hospital, and cared for; and seldom is it but what the indisposition is cured, and she takes her place back again as well as ever. At evening the boys go the rounds to gather up the proceeds of the day's labor, which will average two hundred dozen per day the year through.

WHEN THE KILLING TIME TAKES PLACE.

"Killing time" takes place twice during the year—in the spring, and again at the commencement of the rainy season. All the early chickens are thus disposed of at good prices; and the two-year-old fowl decapitated, to give room for the younger broods, as they are supposed to be past profitable service after the second year.

THE PROFITS FROM ONE YEAR'S BUSINESS

amounted to eleven thousand dollars. The sales were seventy-two thousand dozen of eggs, and nearly twenty thousand chickens and two-year-olds. Mr. SAN FUENTES expresses himself as being perfectly satisfied with the result obtained, and intends to double his stock each year, until every two hundred feet of his extensive farm has its house of fifty tenants.

APPENDIX.

THE ENGLISH STANDARD OF EXCELLENCE.

[FROM TEGETMEIER'S POULTRY BOOK.]

COCHINS.
GENERAL SHAPE.

THE COCK.

Comb—Single, fine, rather small, perfectly straight and upright, with well-defined serrations, and quite free from side-springs.
Beak—Curved, stout at the base and tapering to the point.
Head—Small for the size of the bird and carried rather forward.
Eye—Very bright and clear.
Deaf-ear—Large and pendant.
Wattles—Large, well rounded on the lower edge.
Neck—Hackle very full and abundant, the lower part reaching well on to the back, so as to produce a gradual slant from near the head to the middle of the back.
Back—Broad, with a gentle rise from the middle to the tail; saddle feathers very abundant.
Wings—Very small; the primaries doubled well under the secondaries, so as to be quite out of sight when the wing is closed.
Tail — Very small; the curved feathers numerous, broad, glossy, and soft; the whole tail forming a small hunch, carried rather horizontally than upright.
Breast—Deep, broad, and full.
Thighs—Very large and strong; plentifully covered with perfectly soft feathers, which on the lower part should be curved inward round the hock, so as nearly to hide the joint from view; Falcon or Vulture hocks, that is, those with hard, stiff feathers projecting in a straight line beyond the joint, are objectionable, but not a disqualification.
Fluff—Very abundant and soft, covering the hind parts, and standing out about the thighs.
Legs—Rather short; very thick and bony, wide apart, well feathered on the outside to the toes.
Toes—Straight and strong; the outer and middle toes being well feathered.
Carriage—Not so upright as other breeds, with a contented, intelligent appearance.

THE HEN.

Comb—Single, very small, fine, low in front, erect and perfectly straight; with small, well-defined serrations.
Beak—Small, curved, and tapering.
Head—Very small, neat, and taper.
Eye—Very bright and clear.
Deaf-ear—Rather large.
Wattles—Small, neatly rounded on the lower edge.
Neck — Short; carried forward, the lower part very full and broad; the feathers reaching well on to the back.
Back—Broad, with abundance of soft feathers rising from the middle of the back to the tail.
Wings—Very small; primaries doubled well under the secondaries, so as to be quite out

of sight when the wing is closed; bow of the wings neatly covered by the breast feathers, and the points sunk well into the fluff.

Tail—Very short and small; carried horizontally, and almost hidden in soft feathers.

Breast—Broad and full; carried low.

Thighs—Large; abundantly covered with soft fluffy feathers; curving inward round the hock, so as to nearly hide the joint from view; Vulture or Falcon hocks are objectionable, but not a disqualification.

Fluff—Very soft and abundant, covering the hind parts and standing out about the thighs, giving the bird a very deep and broad appearance behind.

Legs—Short, thick, and bony; standing wide apart; and well feathered on the outside to the toes.

Toes—Strong and straight, the outer and middle toes well feathered.

Carriage—Low, with a contented, intelligent appearance.

BUFF COCHINS.
COLOR OF COCK.

Comb, Face, Deaf-ear, and Wattles—Brilliant red.

Head—Rich, clear buff.

Hackle, Back, Wings, and Saddle—Rich, deep golden buff; the more uniform and even in color the better; quite free from mealiness on the wing.

Breast, Thighs, and Fluff—Uniform clear, deep buff; as free from mottling or shading as possible.

Tail—Rich, dark chestnut, or bronzy chestnut mixed with black. Dark chestnut preferable.

Legs—Bright yellow; feathers clear, deep buff.

COLOR OF HEN.

Comb, Face, Deaf-ear, and Wattles—Brilliant red.

Plumage—Uniform clear, deep buff throughout; the more uniformly clear and free from mottling or shading the better. A clear hackle preferred, but a slight marking at the end of the feathers of the neck not a disqualification.

Legs—Bright yellow, with feathers same color as body feathers.

LEMON COCHINS.
COLOR OF COCK.

Comb, Face, Deaf-ear, and Wattles—Brilliant red.

Head—Lemon, or light orange buff.

Hackle, Back, Wings, and Saddle—Rich, light orange buff; the more uniformly clear and even in color the better, as free as possible from mealy tinge on the wings.

Breast, Thighs, and Fluff—Clear, uniform lemon buff.

Tail—Rich chestnut.

Legs—Bright yellow; feathers lemon buff.

COLOR OF HEN.

Comb, Face, Deaf-ear, and Wattles—Brilliant red.

Plumage—Clear, lemon buff; uniform and even in color throughout, and perfectly free from being mottled or shaded in any part.

Legs—Bright yellow, with feathers same color as body feathers.

In Buff and Lemon Cochins the colors may be either as above, or intermediate betwixt the two; but the colors must be even and as free from mottling or shading as possible. The birds must also match in the pen.

SILVER BUFF COCHINS.
COLOR OF COCK.

Comb, Face, Deaf-ear, and Wattles—Brilliant read.

Head—Light, silvery buff.

Hackle—Rich, gold color.

Back, Shoulder Coverts, and Wings—Bright silvery buff; the more even and uniform in color the better.

Saddle—Rich, gold color.

Breast, Thighs, and Fluff—Clear, light silvery buff.

Tail—Light chestnut; a slight mixture of white not very objectionable, though not desirable.

Legs—Bright yellow, with silvery buff feathers.

COLOR OF HEN.

Comb, Face, Deaf-ear, and Wattles—Brilliant red.

Hackle—Rich, gold color.
Remainder of the Plumage—Clear, light, silvery buff; the more even and uniform in color the better.
Legs—Bright yellow, with feathers same color as body feathers.

SILVER CINNAMON COCHINS.
COLOR OF COCK.

Comb, Face, Deaf-ear, and Wattles—Brilliant red.
Head—Pale light cinnamon.
Hackle—Cinnamon, or rich bright cinnamon, slightly striped with white.
Back, Shoulder, and Wings—Pale buff, or rich bright cinnamon, mixed with white.
Saddle—Light cinnamon, or rich bright cinnamon, slightly striped with white.
Breast, Thighs, and Fluff—Pale buff.
Tail—Rich light cinnamon, or rich bright cinnamon, mixed with white.
Legs—Bright yellow, with feathers of a pale buff color.

COLOR OF HEN.

Comb, Face, Deaf-ear, and Wattles—Brilliant red.
Hackle—Rich deep crimson or chocolate.
Remainder of the Plumage—Pale buff; the more uniform and even in color the better.
Legs—Bright yellow; feathers same color as body feathers.

CINNAMON COCHINS.
COLOR OF COCK.

Comb, Face, Deaf-ear and Wattles—Brilliant red.
Head, Hackle, Back, Wings, and Saddle—Rich dark reddish cinnamon; the more uniform and even in color the better.
Breast, Thighs, and Fluff—The color of wetted cinnamon.
Tail—Rich bronzy black, the lesser coverts edged with very dark reddish cinnamon.
Legs—Bright yellow; with feathers color of the breast feathers.

COLOR OF HEN.

Comb, Face, Deaf-ear, and Wattles—Brilliant red.

Plumage—The color of wetted cinnamon or deep chocolate throughout; the more uniform in color and free from being mottled the better.
Legs—Bright yellow, with feather same color as body feathers.

Value of Points in Buff, Lemon, Silver Buff, Silver Cinnamon and Cinnamon Cochins.

Size	3
Color	4
Head and Comb	1
Carriage of Wings	1
Legs	1
Fluff	1
General Symmetry	2
Condition	2
	15

Disqualification in Buff, Lemon, Silver Buff, Silver Cinnamon and Cinnamon Cochins.

Birds not matching in the pen, or with primary wing feathers, twisted or turned outside the wing, twisted combs, crooked backs, birds without feathers on the legs, or legs of any other color than yellow.

GROUSE OR PARTRIDGE COCHINS.
COLOR OF COCK.

Comb, Face, Deaf-ear, and Wattles—Rich brilliant red.
Head—Rich red.
Hackle—Rich bright red, with a rich black stripe down the middle of each feather.
Back and Shoulder Coverts—Rich dark red.
Wing Bow—Rich dark red.
" *Greater and Lesser Coverts*—Metallic greenish black, forming a wide bar across the wings.
" *Primary Quills*—Bay on outside web, dark on inside web.
" *Secondary Quills*—Rich bay on the outside web, black on the inner web, with a metallic black end to each feather.
Saddle—Rich bright red, with a black stripe down the middle of each feather.
Breast, Under part of Body, and Thighs—Rich deep black.
Tail—Glossy black (white at the base of the feathers objectionable, but not a disqualification.)

Legs—Dusky yellow, with black feathers.

GROUSE COCHINS.
COLOR OF HEN.

Comb, Face, Deaf-ear, and Wattles—Brilliant red.
Head—Rich brown.
Neck—Rich reddish gold color, with a broad black stripe down the middle of the feathers.
Legs — Dusky yellow, with feathers same color as body feathers.

PARTRIDGE COCHINS.
COLOR OF HEN.

Comb, Face, Deaf-ear, and Wattles—Brilliant red.
Neck—Bright gold color on the edge of the feathers, with a broad black stripe down the middle.
Remainder of the Plumage—Light brown distinctly penciled with dark brown; the penciling to reach well up the front of the breast. The shaft of the feathers on the back, shoulder coverts, bow of the wing, and sides, creamy white.
Remainder of the Plumage — Rich brown distinctly penciled with darker brown; the penciling reaching well up the front of the breast, and following the outline of the feathers.
Legs—Dusky yellow, with brown feathers.

Points in Grouse and Partridge Cochins.

Size	3
Black Breast, Thighs, Fluff, and Leg feathers in the Cock	2
Breast of the Hen. Distinctly penciled up the front	
Color of the remaining plumage	2
Head and Comb	1
Carriage of Wings	1
Legs	1
Fluff	1
Symmetry	2
Condition	2
	15

Disqualifications in Grouse or Partridge Cochins.

Birds not matching in the pen—cocks with mottled breasts, hens with pale buff or clay breasts without penciling, twisted combs, flight feathers turned outside the wing, crooked back; absence of feathers on the legs.

WHITE COCHINS.
COLOR OF COCK AND HEN.

Comb, Face, Deaf-ear, and Wattles—Brilliant red.
Plumage — Pure white throughout. The cock as free from yellow tinge as possible.
Legs—Bright yellow.

BLACK COCHINS.
COLOR OF COCK AND HEN.

Comb, Face, Deaf-ear, and Wattles—Brilliant red.
Plumage—Perfectly black throughout. The cock as free from coppery red or brassy color as possible.
Legs—Dark, with yellow tinge and black feathers.

Points in White or Black Cochins.

Size	3
Color of Plumage—Purity of white in the whites, and richness of black in the blacks	4
Head and Comb	1
Carriage of Wings	1
Legs	1
Fluff	1
Symmetry	2
Condition	2
	15

Disqualifications in White or Black Cochins.

Twisted combs, crooked backs, flight feathers turned outside the wing. Birds not feathered on the legs, scales on the legs of the whites either green or willow.

BRAHMAS.
GENERAL SHAPE.
THE COCK.

Beak—Very strong, taper and well curved.
Comb—Pea, small, low in front and firm on the head without falling over to either

side, distinctly divided so as to have the appearance of three small combs joined together in the lower part and back, the largest in the middle, each part slightly and evenly serrated.

Head—Small and slender.

Eye—Prominent and bright.

Deaf-ear—Large and pendant.

Wattles—Small, well rounded on the lower edge.

Neck—Long, neatly curved, slender near the head, the juncture very distinct, hackle full and abundant, flowing well over the shoulders.

Breast—Very full, broad, and round; carried well forward.

Back—Short, broad, flat betwixt the shoulders, saddle feathers very abundant.

Wings—Small; the primaries doubled well under the secondaries, the points covered by the saddle feathers.

Tail—Small; carried very upright, the higher feathers spreading out laterally.

Tail Coverts—Broad, very abundant, soft, and curved over the tail.

Thighs—Very large and strong; abundantly covered with very soft fluffy feathers, curving inward round the hock so as to hide the joint from view. Vulture hocks are objectionable, but not a disqualification.

Fluff—Very abundant and soft, covering the hind parts, and standing out about the thighs, giving the bird a very broad and deep appearance behind.

Legs—Rather short, strong, and bony; standing well apart, very abundantly feathered down the outside to the end of the toes.

Toes—Straight and strong; the outer and middle toe being abundantly feathered.

Carriage—Very upright and strutting.

THE HEN.

Beak—Strong, curved, and taper.

Comb—Pea, very small and low, placed in front of the head, and having the appearance of three very small serrated combs pressed together, the largest in the middle.

Head—Small and slender.

Eye—Prominent and bright.

Deaf-ear—Large and pendant.

Wattles—Small, rounded on the lower edge.

Neck—Rather short, neatly curved, slender near the head, the juncture very distinct, full and broad in the lower part; the feathers reaching well on to the shoulders.

Breast—Very deep, round, broad, and prominent.

Back—Broad and short; the feathers of the neck reaching to betwixt the shoulders, and abundance of soft, broad feathers rising to the tail.

Wings—Small; the bow covered by the breast feathers, the primaries doubled well under the secondaries, the points of the wings clipped well into the abundance of soft feathers and fluff.

Tail—Small; very upright, almost buried in the soft rump feathers.

Thighs—Strong and well covered with very soft feathers, curving round the hock so as to hide the joint from view. Vulture hocks are objectionable, but not a disqualification.

Fluff—Very abundant and soft, standing out about the hind part and thighs, giving the bird a very broad and deep appearance behind.

Legs—Short, very strong, wide apart, abundantly feathered on the outside to the toes.

Toes—Straight and strong, the outer and middle toe being well feathered.

Carriage—Low in comparison to the cock.

PENCILED BRAHMAS.

COLOR OF COCK.

Comb, Face, Deaf-ear, and Wattles—Bright red.

Head—White.

Neck, Hackle—Silvery white, striped with black.

Breast, Underpart of Body, and Thighs—Black, slightly mottled with white.

Back and Shoulder Coverts—Silvery white.

Saddle—Silvery white, striped with black.

Wing Bow—Silvery white.

" *Greater and Lesser Wing Coverts*—Metallic green black, forming a wide well-defined bar across the wing.

" *Secondaries*—White on the outside web, black on the inside web, large green black spot on the end of the feather.

" *Primaries*—Narrow edging of white on the outside web, black on the inside web.

Tail—Black.
Tail Coverts—Rich green black, lesser coverts edged with white.
Legs—Scales yellow, feathers black, mottled with white.

COLOR OF HEN.

Comb, Face, Deaf-ear, and Wattles—Rich bright red.
Head—Gray.
Neck—Silvery white, striped with black.
Remainder of the Plumage—Dull white, minutely and distinctly penciled throughout with dark penciling, so close as almost to cover the ground color, the penciling reaching well up the front of the breast.
Legs—Scales yellow, with a dusky shade.

LIGHT BRAHMAS.
COLOR OF COCK.

Comb, Face, Deaf-ear, and Wattles—Rich bright red.
Head—White.
Neck—White with a distinct black stripe down the center of the feather.
Breast, Underpart of Body and Thighs—White.
Back and Shoulder Coverts—White.
Saddle—White, striped with black.
Wing Bow and Coverts—White.
" *Primaries*—Black.
" *Secondaries*—White on outside web, black on inside web.
Tail—Black.
Tail Coverts—Glossy green black; lesser coverts silvered on the edge.
Legs—Scales bright yellow; feathers white, slightly mottled with black.

COLOR OF HEN.

Comb, Face, Deaf-ear, and Wattles—Bright red.
Head—White.
Neck—White, distinctly striped down the middle of each feather with rich black.
Breast and Back—White.
Wing—White, the primaries alone being black.
Tail—Black, the two highest or deck-feathers edged with white.
Thighs and Fluff—White.

Legs—Bright rich yellow; feathers white, slightly mottled with black.

Points in Brahmas.

Size	3
Color	4
Head and Comb	1
Wings, primaries well tucked under secondaries	1
Legs and featherings of ditto	1
Fluff	1
Symmetry	2
Condition	2
	15

Disqualifications.

Birds not matching in the pen, combs not uniform in the pen, or falling over to one side, crooked backs, legs not feathered to the toes, or of any other color except yellow, or dusky yellow.

MALAYS.
GENERAL SHAPE.
THE COCK.

Beak—Very strong and curved.
Comb—Small, placed quite in front of the head, low and flat, covered over with very small warty indentations.
Head—Long, flat on the top, projecting over the eyes.
Eye—Bright, sunk beneath a projecting eyebrow, the eyelids pearled round the edge.
Face—Very naked and skinny, with a harsh cruel expression.
Wattles—Very small, mere folds of the naked skin of the throat.
Throat—Very skinny, and quite destitute of feathers.
Neck—Very long, slightly curved, rapidly slanting from the head; the hackle very hard, short, and scanty, particularly in the lower part.
Back—Very long, slightly curved, and rapidly slanting from the shoulders to the tail, the shoulder coverts and saddle feathers very short and hard.
Body—Long and round, the feathers on the lower part very short, giving the bird a cut-out appearance.
Wings—Very strong, projecting out prominently from the body even when closed.

APPENDIX.

Breast—Very deep.
Tail—Small, drooping, sickle and tail coverts slightly curved. (The neck, back, and tail forming three slight nearly equal curves.)
Thighs—Very long, round, strong and upright, the feathers very hard, short, and close, the hock joint being bare.
Legs—Very long, strong, round, straight and clean, perfectly free from feathers.
Toes—Very long, straight, strong, and powerful.
Plumage—Very hard, short, close and glossy.
Carriage—Very upright, and tall.

THE HEN.

Beak—Very strong and curved.
Comb—Very small, low, and flat, placed on the front of the head, covered over with small warty indentations.
Head—Long, very snaky, and flat on the top.
Eye—Bright, sunk beneath a projecting eyebrow, eyelids pearled round the edge.
Face—Very naked and skinny, with a cruel expression.
Wattles—Mere folds of the naked skin and throat.
Throat—Quite naked and very skinny.
Neck—Very long, rapidly slanting from the head; neck feathers very hard, short, and close, particularly in the lower part.
Back—Long, rapidly slanting in the tail, shoulder coverts very short.
Body—Long and round, narrow at the insertion of the tail.
Breast—Very deep.
Wings—Very strong, projecting very prominently from the body when closed.
Tail—Small, and carried upright.
Thighs—Very long, strong, and upright; feathers very close and short, the hock joint being nearly naked.
Legs—Very long, clean, straight, round and strong.
Toes—Long, powerful, straight and strong.
Plumage—Very short, hard, close and glossy.
Carriage—Very upright.

COLOR OF MALAY COCK.

Beak—Yellow.
Comb, Face and naked skin of the Throat—Rich bright red.
Eyes—Bright fiery red.
Head and Neck—Rich glossy dark red.
Back and Shoulder Coverts—Glossy red maroon.
Breast—Black, slightly mottled with reddish brown.
Wing Bow—Glossy reddish maroon.
" *Coverts*—Rich metallic greenish or bluish black, forming a wide bar across the wing.
Wing flights—Rich dark red.
Saddle—Rich glossy dark red.
Tail—Rich Green black.
Thighs—Rich black, slightly mottled with reddish brown.
Legs—Bright rich yellow.

COLOR OF MALAY HEN.

Beak—Yellow.
Comb, Face, and Throat—Bright red.
Eyes—Bright fiery red.
Head—Reddish brown.
Neck—Rich glossy reddish brown.
Back and Shoulder Coverts—Rich glossy reddish brown or cinnamon.
Breast and Thighs—Reddish brown or cinnamon.
Wings—Rich glossy reddish brown or cinnamon.
Tail—Rich dark reddish brown.
Legs—Bright rich yellow.

WHITE MALAYS.

Comb, Face, and Naked skin on the Throat—Bright red.
Beak—Bright rich yellow.
Plumage—Pure white throughout.
Legs—Bright rich yellow, yellowish willow permissible.

Points in Malays.

Hight	3
Shortness, hardness, and closeness of plumage	3
Head	1
Color	3
Symmetry	3
Condition	2
	15

Disqualifications in Malays.

Birds not matching in the pen; in the

dark birds legs of any other color except yellow.

DORKINGS.

GENERAL SHAPE.

THE COCK.

Beak—Rather short and stout.
Comb—Either single or rose; if single, erect, straight, serrated, free from side-sprigs; if rose-combed, square in front, straight on the head, without hollow in the middle, large peak behind, inclining very slightly upwards.
Head—Neat.
Wattles—Broad, stout, rounded on the lower edge.
Neck—Very taper and well hackled.
Breast—Very deep, broad and full. Breastbone long.
Body—Large, deep, compact, and plump, the back, belly, breast, and behind, almost forming a square.
Back—Very broad.
Wings—Large.
Tail—Very large, expanded, feathers broad and carried well up.
Sickle Feathers and Tail Coverts—Long, broad, sound and well arched.
Thighs—Short, stout and straight.
Legs—Straight, short, stout, clean, and perfectly free from feathers, spurred on the inside.
Feet—Five-toed, the extra or supernumerary toe well-developed, distinctly separated from the others, and pointing upwards.
Carriage and Appearance—Noble, bulky and grand.

THE HEN.

Beak—Rather short.
Comb—If single, to be well developed, and falling over one side of the face; if rose, square in front, straight on the head, peak behind, inclining slightly upwards.
Wattles—Broad, rounded on the lower edge.
Head—Neat.
Neck—Short and taper.
Breast—Very deep, broad, and full.
Body—Large, compact, plump, and deep.
Back—Broad.
Wings—Large.
Tail—Large, expanded, the feathers broad.
Thighs—Short and stout.
Legs—Short, straight, thick, and strong.
Feet—Five-toed, the extra toe well developed, distinctly separated from the others and inclining upwards.
Carriage and Appearance—Bulky.

SILVER GRAY DORKINGS.

COLOR OF COCK.

Head and Neck Hackle—Clear white.
Comb, Face, and Wattles—Bright red.
Breast, Underpart of Body, and Thighs—Rich glossy black.
Back and Shoulder Coverts—Silvery white.
Saddle—Clear white.
Wing Bow—Silvery white.
  " *Coverts*—Metallic green black, forming a wide bar across the wing.
  " *Primaries*—White on the outside edge of the outer web, black on the inside web.
  " *Secondaries*—Clear white on the outside web, black on the inside web, and also on the end of the feather.
Tail—Rich black.
Sickle Feathers—Rich metallic green black.
Tail Coverts—Rich metallic green black, the lesser ones silvered on the edge.
Legs—White, with a flesh-colored tinge betwixt the scales.

COLOR OF HEN.

Head—Silvery or ashy gray.
Comb, Face, and Wattles—Bright red.
Neck—Silvery white, striped with black.
Breast—Salmon red, shading off to gray towards the thighs.
Back and Shoulder Coverts—Silvery or slaty gray, free from dark bars or marks across the feathers, shaft of feathers white.
Wing Bow—Silvery or slaty gray, shaft of feathers white. Any tendency to red on the wings is highly objectionable.
Coverts and Flights—Slaty gray.
Tail—Dark gray, inside approaching black.
Thighs—Ashy gray.
Legs—White, with a flesh colored tinge betwixt the scales.

APPENDIX.

Points in Silver Gray Dorkings.

Size	3
Color	3
Head and Comb	2
Legs, Feet, and toes	2
Symmetry	3
Condition	2
	15

Disqualifications.

Birds without the fifth toe, or with crooked backs, wry tails, combs not uniform in the pen, white in cock's breast or tail, legs of any color except white.

COLORED DORKINGS.

The color in these not material, providing the birds match in the pen.

Points in Colored Dorkings.

Size	5
Head and Comb	2
Legs, Feet, and Toes	2
Symmetry	4
Condition	2
	15

Disqualifications.

Birds without the fifth toe, or with crooked backs, wry tails, combs not matching in the pen, legs of any other color except white.

WHITE DORKINGS.

Comb, Face, and Wattles—Rich red.

The whole of the plumage in both cock and hen pure white, the more free from yellow tinge the better.

Legs—White.

Points in White Dorkings.

Size	4
Purity of White Plumage	2
Head and Comb	2
Legs, Feet, and Toes	2
Symmetry	3
Condition	2
	15

Disqualifications.

Birds without the fifth toe, or with crooked backs or wry tails, combs not uniform in the pen, colored feathers in any part of the plumage.

SPANISH.

GENERAL SHAPE.

THE COCK.

Beak—Dark horn color, rather long and stout.

Comb—Bright red, large, single, stiff, erect, straight, free from twists in front or falling over to either side at the back, deeply serrated, rising from the beak betwixt the fore part of the nostrils, and extending in an arched form over the back of the head, free from excrescences or side-sprigs, and not of too great thickness at the edge.

Head—Long, broad, and deep-sided.

Eyes—Large, the sight perfectly free, and not obstructed by the white.

Face—Pure opaque white, long and deep, the greater breadth of surface the better, providing it is smooth, free from wrinkles, and the sight not obstructed, rising well over the eye towards the comb in an arched form, extending towards the back of the head, and also to the base of the beak, covering the cheeks, and joining the ear-lobes and wattles.

Ear-lobes—Pure opaque white, very large and pendant, rather thin, smooth, well expanded and free from folds or wrinkles, extending well on each side of the neck, hanging down very low, not pointed, but regularly rounded in the lower part, and meeting in front, behind the wattles.

Wattles—Bright red; very long, thin, ribbon-like, and pendulous; the inside of the upper part, and skin betwixt, white.

Neck—Long, well hackled.

Breast—Round, full, and prominent.

Back—Slanting down to the tail.

Body—Wedge-shaped, narrowing to the tail.

Wings—Large, carried well up to the body.

Tail—Large, expanded and rather upright, but not carried over the back, or squirrel tailed.

Sickle Feathers—Large and well curved.

Thighs—Long and slender.

Legs—Long, dark leaden blue, or blue.

Plumage—Rich glossy black, having a metallic green luster on the hackle, back, wings, saddle, tail coverts, and sickle feathers.

Carriage—Upright and striking.

Disqualifications in Spanish Cocks.

Comb—Falling over to one side, or twisted in front over the nostrils.

Face—So puffy as to obstruct the sight; decided red mark above the eye.

Plumage—Of any other color except black, or metallic green black.

Legs—Of any other color except dark leaden blue, or blue.

THE HEN.

Beak—Dark horn color, long.

Comb—Glossy bright red, large, single serrated, drooping over to one side of the face, free from side-sprigs or duplicature.

Head—Long and deep.

Eyes—Large.

Face—Pure opaque white, smooth and free from wrinkles, with great breadth of surface, rising well over the eye in an arched form, extending well towards the back of the head, and also to the beak, covering the cheek, and joining the ear-lobes and wattles.

Ear-lobes—Pure opaque white, large, pendant, smooth, well expanded, free from wrinkles, regularly rounded on the lower edge.

Wattles—Bright red, thin, pendant, and rounded on the lower edge.

Neck—Long and graceful.

Breast—Round and full.

Back—Slanting down to the tail.

Wings—Ample, carried close up to the body.

Tail—Large, carried rather upright, but not over the back, the two highest feathers slightly curved, especially in pullets.

Thighs—Long and slender.

Legs—Long, dark leaden blue, or blue.

Plumage—Black, with a rich metallic luster on the back and wings.

Disqualifications in Spanish Hens.

Duplicature of comb, comb small and erect (prick-combed); decided **red mark** over the eye; plumage of any other color except black, or metallic black; legs of any other color except dark leaden blue, or blue; birds that are trimmed in any part whatever.

Points in Spanish Fowl.

Comb.................... 2
Face..................... 3
Ear-lobe................. 3
Purity of White, Face and Ear-lobe............... 2
Symmetry................ 3
Condition of Plumage..... 2

15

Carriage—Upright, movement quick.

GAME.

GENERAL SHAPE.

THE COCK.

Beak—Strong, curved, very stout at the base.

Comb—In a chicken that has not been dubbed, single, small and thin, low in front, serrated, erect, and straight; in older birds, neatly dubbed, smooth, and free from warty appearances, small feathers or ridges on the edges.

Head—Long, thin, and taper, very strong at the juncture with the neck.

Eyes—Large, bright, and prominent, perfectly alike in color, with a quick, fearless expression.

Face and Throat—Lean and thin.

Neck—Rather long and neatly arched, hackle short and very close.

Back—Rather short, flat, broad across the shoulders and narrowing to the tail.

Breast—Broad, round, and full.

Stern—Slender and very neat, saddle feathers very short and close.

Wings—Strong, long, and very powerful; the butts and shoulder part slightly raised, as if for a sudden spring, the remainder of the wings not drooping, but carried neatly and compactly to the sides, passing over the upper part of the thighs, the points resting under the saddle feathers.

Tail—Rather long, the feathers very sound and not too broad, carried well together, and not spread out, scattered or loose.

Sickle Feathers and Tail Coverts—Perfectly sound, narrow, hard and wiry, not hanging loosely, well carried and neatly curved;

the whole of the tail going backwards and not upright over the back, or squirrel-tailed.

Thighs—Round, stout, hard, and firm, rather short in proportion to the shank, placed well up towards the shoulders, and covered with very close short feathers, so as to have a velvety appearance.

Legs—Rather long, strong, bony, clean, standing well and evenly apart; the spurs set on low; the scales close and smooth.

Feet—Broad, flat, and thin; toes long, spreading, and straight, well furnished with strong nails, with the hind toe set low on the foot, standing well backwards and flat on the ground, not merely touching with the point of the toe, or duck-footed.

Plumage—Close, sleek, and glossy, body feathers short, hard and firm, quills very strong.

Body in hand—Very muscular, and firm, not soft or hollow on the sides, perfectly straight in the breast and back, and quite even in the hip bones.

Carriage—Upright, active and quick.

THE HEN.

Beak—Long, slightly curved, sharp at the point, and stout at the base.

Comb—Single, small, and thin, low in front, evenly serrated, perfectly erect and straight.

Head—Long, slender, very neat and taper.

Eyes—Bright, large, and prominent, perfectly alike in color, with a quick and fiery expression.

Face—Lean and thin.

Deaf-ear—Very small, and close to the face.

Wattles—Small, thin, and neatly rounded on the edge.

Throat—Neat, the feathers very short and close.

Neck—Long, feathers very short, giving the neck a slender and very graceful appearance.

Back—Moderate in length, perfectly flat and broad across the shoulders, and narrowing to the tail.

Wings—Long and powerful, the butts and shoulders carried rather high, so as to cause a perfectly flat back, the points not drooping, but carried compactly to the sides.

Tail—Moderate in length, not carried over the back, but extending backwards; the feathers not scattered or spread out, but held neatly together.

Breast—Broad, round, and prominent.

Thighs—Stout, round, and neat, the feathers short and very close.

Legs—Long, very bony, clean, and taper, the scales narrow, smooth, close, and neat.

Feet—Broad, flat, and thin; toes spreading, long and straight, well furnished with strong nails, the hind toe set low on the foot, standing well backwards, and not duck-footed.

Plumage—Very close, sleek, and glossy; body feathers short, hard and firm, quills strong.

Carriage—Rather upright, very neat, quick, and active.

BLACK-BREASTED RED GAME.

COLOR OF COCK.

Head—Very rich dark red.

Comb, Face, and Jaws—Very bright red.

Eyes—Bright, clear, deep bay.

Neck Hackle—Rich red, free from black or dark stripes.

Back, Shoulder, and Shoulder Coverts—Rich dark red.

Wing Butts—Black.
" *Bow*—Rich dark red, perfectly free from black feathers.
" *Greater and Lesser Coverts*—Metallic green black, forming a wide bar across the wing, perfectly even, well defined, and not irregular on the edges.
" *Primaries*—Bay on the outside web, black on the inside.
" *Secondaries*—Rich clear bright bay on the outside web, black on the inside web, with a rich metallic green black spot on the end of the feather.

Saddle—Rich red.

Tail—Rich black.

Sickle Feathers and Tail Coverts—Very rich metallic green black.

Breast, Underpart of body, and Thighs—Rich

black, perfectly free from any admixture of red or other color.

Legs—Either willow, olive, yellow, white, or blue. The colors preferred in the order in which they are named.

COLOR OF HEN.

Head—Brown.
Comb, Face, Deaf-ear, and Wattles—Very bright red.
Neck—Light brownish yellow, striped with black.
Back and Shoulder Coverts—Brown.
Wing, Bow, Shoulder, and Coverts—Same color as back, perfectly free from red.
" *Primaries and Secondaries*—Brown.
Tail—Dark brown, approaching black.
Breast—Deep salmon, shading off to ashy brown towards the thighs.
Thighs—Ashy brown.
Legs—To match those of the cock.

BROWN RED GAME.

COLOR OF COCK.

Head—Very dark red.
Comb, Face, and Jaws—Bright red or dark purple (gipsy-faced.)
Eyes—Dark brown or black.
Neck Hackle—Dark red, shaft of feathers black.
Back and Shoulder Coverts—Dark crimson red.
Saddle—Dark red, shaft of feathers black.
Wing, Butts—Black or very dark dusky brown.
" *Shoulder and Bow*—Dark crimson red.
" *Coverts*—Rich glossy black
" *Primaries*—Dusky black.
" *Secondaries*—Black, with a metallic luster towards the end of the feathers.
Tail—Black.
Sickle Feathers and Tail Coverts—Rich glossy black.
Breast—Reddish brown streaked with black, shaft of feathers black; the ground color becoming darker as it approaches the lower part and thighs.

Thighs—Dusky black.
Legs—Olive bronzy black, or dark willow.

COLOR OF HEN.

Head—Dark dusky brown, approaching a dusky black.
Comb, Face, Deaf-ear, and Wattles—Bright red or dark purple.
Eyes—Very dark brown or black.
Neck—Coppery yellow, striped with black.
Remainder of the Plumage—Very dark brown, approaching to black.
Legs—To match those of the cock.

GINGER RED GAME.

COLOR OF COCK.

Head—Red.
Face and Jaws—Reddish purple.
Eyes—Brown.
Neck Hackle—Rich clear red.
Back, Shoulder Coverts and Bow of the Wings Rich red.
Wing—Primaries and secondaries brownish red.
Saddle—Rich clear red.
Tail—Black.
Sickle Feathers and Tail Coverts—Rich black, the lesser coverts edged with red.
Breast—Ginger red, becoming darker as it approaches the thighs.
Thighs—Dusky red.
Legs—Olive, bronzy black, or dark willow.

COLOR OF HEN.

Head—Yellowish brown.
Comb, Face, Deaf-ear, and Wattles—Purple.
Eyes—Brown, perfectly alike in color.
Neck—Golden yellow, striped with black.
Breast—Higher part towards the throat yellowish brown, shaft and a narrow margin of the feathers a much lighter shade.
" Lower part and sides, dark dusky brown, with a narrow margin of the feathers of a golden ginger shade.
Remainder of the Plumage—Yellowish brown, with a narrow margin of the feathers of a golden ginger shade.
Legs—Same color as those of the cock.

YELLOW DUCK-WING GAME.

COLOR OF COCK.

Head—Straw-colored yellow.
Comb, Face, and Jaws—Bright red.
Neck Hackle—Clear, straw color, free from black.
Back, Shoulder Coverts, and Bow of the Wings, Rich, uniform, bright copper or maroon; the more even, clear, and unmixed in color the better.
Wing Butts—Black.
" *Greater and Lesser Coverts*—Steel blue, or metallic black, forming a wide bar across the wing.
" *Primaries*—Straw white on the outside web, dark on the inside web.
" *Secondaries*—White on the outside web, black on the inside, and on the end of the feathers.
Saddle—Clear straw color.
Breast, Underpart of Body, and Thighs—Rich black.
Tail—Black.
Sickle Feathers and Tail Coverts—Rich metallic green-black.
Legs—Willow, yellow, or olive.

COLOR OF HEN.

Head—Gray.
Comb, Face, Deaf-ear, and Wattles—Bright red.
Neck—White, striped with black.
Breast—Salmon red, shading off to ashy gray towards the thighs.
Back and Shoulder Coverts—Bluish or slaty gray, shaft of feather white.
Wing, Shoulder, and Bow—Slaty or bluish gray, shaft of feather white. Red or brown on the wing very objectionable.
Tail Coverts and Flights—Slaty or bluish gray.
Tail—Dark gray, the inside approaching black.
Thighs—Ashy gray.
Legs—To match those of the cock.

SILVER DUCK-WING GAME.

COLOR OF COCK.

Head—Silvery white.
Face, Jaws, and Comb—Bright red.
Neck—Hackle clear white, without any mixture of black or other color.
Breast, Underparts of Body, and Thighs—Black.
Back and Shoulder Coverts—Silvery white.
Saddle—Clear white.
Wing Butts—Black.
" *Bow*—Silvery white.
" *Coverts*—Steel blue, forming a wide bar across the wing.
" *Primaries*—White on the outside web, dark on the inside web.
" *Secondaries*—Clear white on the outside web, black on the inside web, and on the end of the feathers.
Tail—Black.
Sickle Feathers and Tail Coverts—Metallic green-black; the lesser tail coverts slightly edged with white.
Legs—Willow, olive, bronze, or blue.

COLOR OF HEN.

Head—Silvery gray.
Comb, Face, Deaf-ear, and Wattles—Bright red.
Neck—Silver, striped with black.
Breast—Salmon.
Back and Shoulder Coverts—Silvery or ashy gray, shaft of feather white.
Wing Bow—Ashy gray, shaft of feather white. Red or brown on the wing very objectionable.
" *Flight and Coverts*—Gray.
Tail—Dark gray, approaching black.
Thighs—Ashy gray.
Legs—To match those of the cock.

BIRCHEN YELLOW GAME.

COLOR OF COCK.

Head—Dark straw color.
Face and Wattles—Either red or purple.
Neck—Hackle, deep straw color, striped with reddish brown.
Breast—Reddish brown, shaft and narrow margin of the feathers cream color.
Back and Shoulder Coverts—Rich coppery straw, marked with reddish brown.
Saddle—Deep straw, striped with reddish brown.
Wing Butts—Dull black.

Wing Bow — Rich dark coppery straw, slightly marked with reddish brown.
" *Coverts*—Cream color, mottled with reddish brown and tipped with chocolate.
" *Flight*—Reddish brown.
Tail—Black.
Sickles—Bronzy black.
Tail Coverts—Bronzy black, the lesser with a narrow margin of cream color.
Legs—Bronzy black, olive, willow, or yellow.

COLOR OF HEN.

Head—Dark gray.
Comb, Face, Deaf-ear, and Wattles—Either red or purple.
Neck—Gray, striped with dull black.
Breast—Grayish brown, shaft and margin of feather creamy white.
Back and Shoulder Coverts—Grayish brown, shaft of feather dull creamy white.
Wing Bow—Grayish brown, shaft of feather dull creamy white.
" *Coverts*—Grayish brown.
" *Flights*—Dark gray.
Tail—Dark grayish brown.
Thighs—Grayish brown.
Legs—To match those of the cock.

PILE GAME.

COLOR OF COCK.

Head—Deep chestnut red.
Comb, &c.—Rich bright red.
Neck Hackle and Saddle—Light chestnut red on the outside of the web of the feather; the middle of each feather white towards the end.
Breast—Higher part marbled red and white, lower part white, or entirely white.
Back, Shoulder Coverts and Bow of the Wings, Rich, uniform red.
Greater and Lesser Wing Coverts—White, edged with red.
Wing Secondaries—White on the outside web, red on the inside web, with a rich red spot on the end of the feather.
" *Primaries*—White.
Thighs—White.
Tail—White.
Legs—Yellow, willow, or white.

COLOR OF HEN.

Comb, Face, Deaf-ear, and Wattles—Bright red.
Neck—Chestnut and white.
Breast—Chestnut red on the front part, mottled with white on the lower part.
Thighs and Tail—White.
Remainder of the Plumage—White, mottled with light chestnut red.
Legs—To match those of the cock.

WHITE GAME.

Comb, Face, Deaf-ear, and Wattles—Very bright red. The whole of the plumage clear white. The cock's plumage as free from yellow tinge as possible.
Legs—Yellow or white.

BLACK GAME.

Comb, &c.—Bright red.
The whole of the plumage glossy black, with a metallic luster on cock's hackle, back, saddle, wings, and tail.
Legs—Bronzy black, dark olive, or leaden black.

Points in Game.

Shape of Head and neck	2
Body and wings	2
Tail	2
Thighs, Legs, and Toes	2
Color of Plumage	3
Symmetry, Handling	2
Condition, and Hardness of plumage	2
	15

Disqualifications.

Color of legs, or plumage, not matching in the pen; crooked backs or breasts; adult cocks not dubbed.

HAMBURGS.

GENERAL SHAPE.

THE COCK.

Beak—Medium.
Comb—Double, not so large as to overhang the eyes or beak, square in front, fitting

close and straight on the head without inclining to either side, no hollow in the center, uniform on each side, the top covered over with small points, with a peak behind, inclining very slightly upwards.

Head—Rather short and small.
Eye—Full and quick.
Deaf-ear—Not pendant, but fitting close to the face, flat, of medium size, round, and even on the surface.
Wattles—Broad, thin, and well rounded on the lower edge.
Neck—Taper, the higher part carried well over the back, hackle full, the lower part flowing well on to the shoulders.
Breast—Round, full, and prominent, carried well forward.
Back—Short, well furnished with saddle feathers.
Wings—Ample, points carried rather low.
Tail—Full, expanded, sickle feathers well curved.
Thighs—Short and neat.
Legs—Slender, rather short, very neat, and taper.
Plumage—Rich and glossy.
Carriage—Upright and strutting, graceful, quick, and restless.

THE HEN.

Beak—Rather small.
Comb—Same shape as that of cock, but very much less; smaller in the penciled than in the spangled varieties.
Head—Small and very neat.
Eye—Full and very quick.
Deaf-ear—Small, flat, rounded in the lower part, fitting closely to the face, and not pendant.
Wattles—Small and thin, rounded on the lower edge.
Neck—Taper and very graceful.
Breast—Broad, plump, and carried forward.
Back—Rather short, but not so much so in appearance as in the cock.
Wings—Ample, carried very neatly to the body.
Tail—Full, expanded, and well carried.
Thighs—Short and neat.
Legs—Very slender, neat and taper.
Plumage—Close and glossy.
Carriage—Graceful, quick, and restless.

GOLDEN PENCILED HAMBURGS.

COLOR OF COCK.

Comb, Face, and Wattles—Rich red.
Deaf-ear—Pure opaque white, free from red on the edge.
Head and Hackle—Clear reddish bay.
Back, Saddle, Bow of the Wing, Shoulder and Wing Coverts—Rich deep reddish bay.
Flight—Reddish bay on the outside web, black on the inside web.
Secondaries—Reddish bay on the outside web, the inside web penciled across with broad black marks, each feather ending with a rich black spot.
Breast and Thighs—Reddish bay.
Tail—Black.
Sickle Feathers and Tail Coverts—Rich black down the middle of the feather, the entire length edged with bronze, each bronze edge as near one-fourth the width of the feather as possible; the more distinct the two colors the better.
Legs—Slaty blue.

COLOR OF HEN.

Comb, Face, and Wattles—Rich red.
Deaf-ear—Pure opaque white, free from red on the edge.
Head and Neck—Clear deep golden bay.
Remainder of the Plumage—Clear deep golden bay, free from either lacing or mossing; each feather (including tail feathers) distinctly penciled across with rich black; the penciling not to follow the outline of the feather, but to go straight across on each side of the shaft. The two colors distinct, well defined, and not shading into each other.
Legs—Slaty blue.

SILVER PENCILED HAMBURGS.

The same standard will apply to the Silver Penciled Hamburgs, substituting a clear silvery white ground for a golden one. The silver cock as free as possible from yellow tinge.

PENCILED HAMBURGS.

Points in Cocks.

Comb	3
Deaf-ear	2
Color of plumage, except tail, sickle feathers, and tail coverts	3
Color of Tail, Sickle Feathers and Tail Coverts	3
Symmetry	2
Condition	2
	15

Points in Hens.

Comb	2
Deaf-ear	2
Purity of Color in Head and Neck	3
Purity of ground color, and accurate and distinct penciling in every part, except head and neck	4
Symmetry	2
Condition	2
	15

Disqualifications.

Hen-feathered cocks, crooked backs, wry tails, combs single or falling over to one side, red deaf-ears, shanks of any other color except blue.

GOLDEN-SPANGLED HAMBURGS.

COLOR OF COCK.

Comb, Face, and Wattles—Rich bright red.
Deaf-ear—Opaque white.
Head—Deep reddish bay.
Hackle—Rich deep golden bay, each feather striped down the centre with rich green black, each color well defined, and not clouded.
Breast, Underpart of Body, and Thighs—Golden bay, free from mossing, streaking, or lacing, each feather ending with a round, large, rich black moon or spangle, the moons increasing in size in proportion to the size of the feather.
Back and Shoulder Coverts—Rich deep reddish bay, distinctly spangled with rich metallic black, the texture of the feather giving the spangle a starry or rayed appearance.
Saddle—Rich reddish golden bay, each feather stripped down the center with rich metallic green black.
Wing Bow—Rich reddish golden bay, distinctly spangled with black.
" *Bars*—The greater and lesser wing coverts clear reddish golden bay, free from lacing, each feather ending with a large round green-black spangle, forming two distinct parallel green-black bars across the wing.
" *Primaries*—Bay, ending with a black spot.
" *Secondaries*—Rich golden bay, each feather ending with a rich green-black spot.
Tail—Black.
Sickle Feathers and Tail Coverts—Rich green-black.
Legs—Slaty blue.

COLOR OF HEN.

Comb, Face, and Wattles—Rich bright red.
Deaf-ear—Opaque white.
Head—Golden bay, distinctly tipped with black.
Neck—Golden bay, each feather distinctly striped down the centre with rich green-black, the colors distinct and not clouded.
Breast, Underpart of Body and Thighs—Clear golden bay, free from mossing or lacing, each feather ending with a distinct large, round, rich green-black moon or spangle, the moons increasing in size in proportion to the size of the feather.
Back, Shoulder Coverts, and Rump—Rich clear golden bay, free from mossing or lacing, each feather ending with a distinct large, round, rich green-black spangle.
Wing Bow—Rich clear golden bay, each feather ending with a distinct round rich green-black spangle.
" *Bars*—Greater and lesser wing coverts rich clear golden bay, free from lacing, each feather ending with a large, round, rich, green-black spangle, forming two distinct parallel green-black bars across the wings.

Wing Primaries—Golden bay, each feather ending with a black spangle.

" *Secondaries*—Golden bay, each feather ending with a rich green-black half moon or crescent-shaped spangle, termed by the Lancashire fanciers, "lacing on the top of the wing above the flight."

Tail—Black.

Tail Coverts—Golden bay, free from mossing or lacing, each feather ending with a rich green-black spangle.

Legs—Slaty blue.

Hens in a pen to match as nearly as possible in size of markings and depth of color.

SILVER-SPANGLED HAMBURGS.

COLOR OF COCK.

Comb, Face, and Wattles—Rich bright red.

Deaf-ear—Opaque white.

Head—Silvery white.

Hackle—Silvery white, free from yellow tinge, the longest feathers ending with a small black spangle.

Breast, Underpart of Body and Thighs—Clear silvery white, free from lacing or mossing, each feather ending with a distinct large, round, rich black moon or spangle, the moons increasing in size in proportion to the size of the feather.

Back and Shoulder Coverts—Pure white, free from yellow tinge, distinctly spangled with black, the texture of the feather giving the spangle a starry or rayed appearance.

Saddle—Silvery white, free from yellow, the largest feathers ending with a small black spangle.

Wing Bow—Pure white, distinctly spangled with black spangles.

" *Bars*—The greater and lesser wing coverts clear silvery white, free from lacing, each feather ending in a large green-black moon or spangle, forming two distinct parallel black bars across the wing.

" *Primaries*—Pure white, each feather ending with a distinct black spangle.

" *Secondaries*—Pure white, each feather ending in a half-moon shaped green-black spot.

Tail—White on the outside, each feather ending in a large black spangle.

Sickle Feathers and Tail Coverts—White, each feather ending with a rich green-black spangle.

Legs—Slaty blue.

COLOR OF HEN.

Comb, Face, and Wattles—Rich bright red.

Deaf-ear—Opaque white.

Head—Silvery white, distinctly spangled with small black spangles.

Neck—Clear silvery white, each feather distinctly striped towards the end with rich black, each color well defined and not clouded.

Breast, Underpart of Body, and Thighs—Clear silvery white, free from lacing or mossing, each feather ending with a distinct large, round black moon or spangle, the moons increasing in size in proportion to the size of the feather.

Back, Shoulder Coverts, and Rump—Clear silvery white, free from mossing or lacing, each feather ending with a distinct large, round, rich green-black moon or spangle.

Wing-Bow—Clear silvery white, each feather ending with a distinct round, rich green-black spangle.

" *Bars*—Greater and lesser wing coverts clear silvery white, free from lacing or mossing, each feather ending with a large round greenish-black spangle, forming two distinct parallel black bars across the wing.

" *Primaries*—White, each feather ending with a distinct black spangle.

" *Secondaries*—Clear silvery white, each feather ending with a large half-moon shaped green-black spangle, termed by the Lancashire fanciers "lacing on the top of the wing."

Tail—White on the outside, each feather ending with a large round black spangle.

Tail Coverts—Clear silvery white, free from mossing or lacing, each feather ending with a distinct large, round, green-black spangle.

Legs—Slaty blue.

Hens in a pen to match as nearly as possible in size of markings and depth of color, &c.

Points in Spangled Hamburg Cocks.

Comb	2
Deaf-ear	2
Colors and Marking of Head, Hackle, Back, Saddle, and Tail	3
Breast, Underparts of Body and Thighs	2
Wings and Bars	2
Symmetry	2
Condition	2
	15

Points in Spangled Hamburg Hens.

Combs	2
Deaf-ear	2
Neck most distinctly and evenly striped	1
Remainder of Plumage (except tail in Golden) clearness of ground color, evenness and distinctness of spangling, with rich large round spangles	4
Bars	2
Symmetry	2
Condition	2
	15

Disqualifications.

Hen-feathered cocks, crooked backs, wry tails, combs single, or falling over to one side, red deaf-ears, birds without distinct bars across the wing. Legs of any other color except blue.

BLACK HAMBURGS.

Comb, Face, and Wattles—Rich bright red, the face perfectly free from white.
Deaf-ear—Pure opaque white; round and small, fitting close to the face; not pendent.
Plumage—Very rich glossy green-black.
Legs—Blue or dark leaden blue.

Points in Black Hamburgs.

Comb, Head, and Face	3
Deaf-ear	2
Plumage	4
Shape	4
Condition	2
	15

Disqualifications.

Combs falling over to one side, or so large as to obstruct the sight, red deaf-ears, crooked backs, wry tails, or legs of any color except blue or dark leaden blue.

POLISH.

GENERAL SHAPE.

THE COCK.

Crest—Composed of feathers similar in texture to the hackle, very large, round, close, and well fitted on the crown of the head, falling backwards, and rather lower on the sides than over the beak, but not so low on the sides as to prevent the bird from seeing.
Head—With round protuberance on the top, concealed by the large crest.
Eye—Large, full, and bright.
Deaf-ear—Small, even on the surface, rounded on the lower edge.
Wattles—In the unbearded varieties, thin and pendulous; in the bearded varieties, none—the underside of the beak and throat being covered with a full, close, muffy beard.
Neck—Medium in length, slightly and neatly curving over the back and well hackled.
Breast—Deep, full, round, and carried prominently forward.
Back—Perfectly straight, wide betwixt the shoulders, and tapering to the tail; hip-bones even.
Wings—Ample.
Tail—Large, rather erect, expanded, and well adorned with sickle feathers.
Thighs—Short in the white-crested black, rather long in the spangled varieties.
Legs—Rather short in the white-crested blacks, long in the spangled varieties.
Carriage—Erect.

THE HEN.

Crest—Very large, round, straight on the head, not inclining to either side, the surface close, firm, and even.
Head—Round, the protuberance concealed by the crest.
Eye—Large, full, and bright.
Deaf-ear—Small, even on the surface, and rounded on the lower edge.

Wattles—In the unbearded varieties, small and thin; in the bearded varieties, none—the throat and underside of the beak being covered with a full close beard.
Neck—Rather short and taper.
Breast—Very full, round, and prominent.
Back—Straight, the hip-bones even.
Wings—Ample.
Tail—Large, expanded, and broad at the end.
Thighs—Short, in the white-crested black, rather long in the spangled varieties.
Legs—Clean, neat, and taper; short in the white-crested blacks, rather long in the spangled varieties.
Carriage—Rather upright.

WHITE-CRESTED BLACK POLISH.

COLOR.

Crest—Pure white; the less black in front the better.
Deaf-ear—Pure opaque white.
Remainder of the Plumage—Uniformly rich glossy black.
Legs—Leaden blue, or black.

Points in White-Crested Black Polish.

Size of Crest	3
Shape of Crest	3
Crest of the purest white, and most free from black	2
Deaf-ear	1
Richest black Plumage	2
Symmetry	2
Condition and General Appearance	2
	15

Disqualifications.

Crooked backs, wry tails, white feathers in any part except the crest, legs of any other color except dark leaden blue, or blue.

GOLDEN-SPANGLED POLISH.

COLOR OF COCK.

Crest—Golden bay, laced with black; in adults, white feathers may appear.
Hackle and Saddle—Golden bay, the end of each feather laced with black.
Breast—Clear golden bay, free from mossing, each feather ending with a round rich black spangle, the spangle increasing in size in proportion to the size of the feather.
Back, Shoulder Coverts, and Bow of the Wing—Rich golden bay, spangled with black, the texture of the feather giving the spangle a rayed appearance.
Bars—Greater and lesser wing coverts, golden bay, each feather laced on the edge with black, and ending with a large black spangle, forming two distinct black bars across the wing.
Primaries—Bay, ending with a black spot.
Secondaries—Golden bay, with a distinct crescent-shaped green-black mark on the end of each feather.
Thighs—Bay, spangled with black.
Tail—Rich golden bay, each feather ending with a rich black spot.
Sickle Feathers—Rich golden bay, ending with a rich black spangle.
Tail Coverts—Rich golden bay, edged with rich black, and ending with a rich black spangle.
Legs—Blue.

COLOR OF HEN.

Crest—Golden bay, each feather laced with black; in adults, white feathers may appear.
Neck—Golden bay, laced with black.
Breast, Underparts of Body and Thighs—Clear golden bay, free from mossing, each feather ending with a distinct round, rich, black spangle, the spangle increasing in size in proportion to the size of the feather.
Back, and Shoulder Coverts—Golden bay, each feather ending with a distinct round black spangle.
Wing Bow—Golden bay, each feather ending with a crescent-shaped black spangle.
Wing Coverts—Golden bay, each feather laced or edged with black, and ending with a large black spangle, forming two distinct black bars across the wing.
Primaries—Bay, each feather ending with a black spot.
Secondaries—Golden bay, each feather ending with a crescent-shaped black mark.
Tail—Bay, each feather ending with a large black spangle.
Legs—Blue.

SILVER-SPANGLED POLISH.

Color and marking the same as in Golden, substituting Silvery White Ground for Golden Bay.

Points in Spangled Polish.

Size of Crest	3
Shape of do	3
Color of do	1
Plumage accurately marked according to the foregoing rules	2
Purity of Ground Color	1
Bars	1
Symmetry	2
Condition	2
	15

Disqualifications.

Crooked Backs, wry tails, legs of any other color except blue.

SULTANS.

THE COCK.

Comb—Composed of hackle feathers, full, and arched over the eyes, and round head, full in center, and falling softly and evenly round at back, not straight and stiff as in Polish; the front free from feathers falling forward, and neatly arched at both sides.

Beak—Brilliant white, tinged with red at base, very curved, and with broad cavernous nostrils.

Comb—Invisible, or two small spikes, brilliantly red.

Muffling—Thick and close round the throat, meeting the crest, and covering the face.

Eye—Bright, vivacious, and intelligent.

Wattles—Small and rather shriveled.

Neck—Rather short, carried well back, very arched, and very thickly hackled.

Breast—Deep, full, round, and carried well forward.

Body—Very square, deep, and carried low.

Back—Straight, and rather broad.

Wings—Ample, and carried down.

Tail—Large, erect, and well sickled.

Thighs—Very short and well feathered.

Legs—Very short, feathered to the toes, with full, long vulture hocks.

Toes—Straight, five in number.

Color of Plumage—Brilliantly white throughout.

Carriage—Rather low, brisk, and vivacious.

THE HEN.

Crest—Full, round, close, and globular.

Eye—Bright and intelligent.

Muffling—Very thick and close round the throat, going well back, covering the face and meeting the crest.

Beak—Curved, clear, transparent white.

Neck—Short, fully arched, and very thickly feathered, carried well back.

Breast—Full, deep, and prominent.

Back—Straight and broad.

Body—Very square, and carried low and forward.

Wings—Full, and carried low.

Tail—Large, erect, and well expanded.

Thighs—Very short, and well feathered.

Legs—Very short, feathered to the toes, with full, large vulture hocks.

Toes—Five in number.

Color of Plumage—Brilliantly white throughout.

Carriage—Low, forward, brisk, and lively.

Points

Crest	4
Muffling	3
Shape	3
Leg-feathering	3
Condition	2
	15

Disqualifications.

Any color but white in the plumage, crooked crest, bare red face, or absence of muffling, deficiency of leg-feathering, or absence of vulture hocks, beak any color but white, deformity of any kind.

HOUDANS.

THE COCK.

Crest—Composed of hackle feathers, full, and well arched, falling back, and right and left of comb, clear of the eye, rather than over it.

Comb—Well developed, large, red, and branching, broad at base, well indented, looking like a mass of coral with antler-like branches, inclining rather backward into the crest.

Beak — Curved, with nostrils wide and cavernous, as in Polish, dark horn color.

Eye—Large, full, bright, and lively; color various.

Wattles—Thin, rather long, neatly rounded, and bright red.

Muffling or Beard—Full and thick under beak, and reaching well back in a curve to the back of eye.

Face—Red, the less seen the better.

Breast—Deep, full, and plump.

Back—Wide and straight.

Wings—Moderate, and carried well up.

Tail—Moderate, erect, and well sickled.

Thighs—The shorter the better.

Legs—Fine in bone, white shaded.

Toes—Five in number, the fifth curved upwards at back.

Color—Broken black and white, as evenly broken as possible, free from colored feathers, which, however, though objectionable, are not a disqualification.

Carriage—Lively, brisk, well set up, and spirited.

THE HEN.

Crest—Large, compact, and even, as in Polish.

Comb—Small, branching, and coral-like.

Eye—Full and bright.

Wattles—Small, red, and neatly rounded.

Muffling—Full, forming a thick beard reaching back to the eye.

Neck—Rather short, full feathered, and arched.

Breast—Full and deep.

Back—Wide and straight.

Wings — Moderate, and carried closely to body.

Tail—Moderate, and fan-like, carried well up.

Thighs—Short.

Legs—Fine in bone, white, or shaded in color.

Toes—Five in number, the hind or fifth claw curved upwards.

Color—As in cock.

Carriage—Brisk and rather upright.

Points.

Size	4
Crest	4
Symmetry	2
Plumage	2
Condition	2
Five Claws	1
	15

Disqualifications.

Absence of crest. Deformity of any kind. Main color or ground color other than black and white.

CREVE CŒURS.

THE COCK.

Crest—As in the Polish cock, but perfectly black; white feathers a defect, but not a disqualification.

Head—As in Polish cock.

Comb—Brilliant red, two-horned in shape but free from tynes, slightly sprigged at base, of good size, showing well in front of the crest.

Eye—Full, bright, and very vivacious.

Deaf-ears—Small and nearly concealed.

Face—Red, well muffled.

Wattles—Moderately pendulous, and evenly rounded, brilliant red.

Muffling—Close and thick, running to back of eye in a handsome curve.

Beak—Black, with horn-colored tip, strong and well curved, with highly arched broad nostrils, as in Polish.

Neck—Moderate in length, thickly hackled, well arched, and carried a little back.

Breast—Broad and full, carried well forward.

Back—Wide, perfectly straight, and free from deformity.

Body—Long and square.

Wings—Closely set, and well clipped up.

Tail—Full and ample, well sickled, and carried rather erect.

Thighs—Rather short, well set in body.

Legs—Black or slate; the shorter the better, rather fine in the bone; free from feathers.

Carriage—Upright, smart, vivacious, and watchful.

Color—Brilliant black. Red or straw feathers in the hackle or saddle undesirable, but not a disqualification.

THE HEN.

Crest—Full and globular, as in the Polish Black; white feathers objectionable, but not a disqualification.
Head—As in Polish.
Eye—Full and bright.
Deaf-ears—Small, hidden by muffling.
Muffling—Thick and full, extending well back to crest, and forming a thick beard under the beak.
Wattles—Very small and neatly rounded.
Neck—Thick and arched.
Breast—Full, plump, and carried well forward.
Body—Square, and carried low.
Back—Straight and broad.
Wings—Well clipped up.
Tail—Large and well expanded.
Thighs—Short, and well set into body.
Legs—Short as possible, free from feathers, rather small in bone, slate or black in color.
Carriage—Upright and vivacious.
Color—Brilliant black; a brown tinge very undesirable.

Points in Creve Cœurs.

Size	4
Crest	3
Shape and Symmetry	2
Color	3
Condition	2
Comb	1
	15

Disqualifications in Creve Cœurs.

Deformity of any kind. Colored feathers elsewhere than in crest, neck, or saddle, feathered legs, and shanks of any other color than black or slate.

LA FLECHE.

THE COCK.

Beak—Black, strong, and curved; nostrils, wide and cavernous, as in Polish, with small spot or knob of bright red flesh at junction of nostril with beak.
Comb—Branching and antler-like, like two horns pointed straight up, brilliant red.
Ear-lobes—Large, and as white as possible.
Head—Long.
Eye—Bright, large, and watchful.
Face—Red, and rather bare.
Wattles—Red, long, and pendulous, well rounded.
Neck—Long, rather curved, and upright; hackle thick, but rather short.
Back—Very long and broad, slanting towards the tail.
Wings—Long, and well clipped in.
Breast—Broad, and rather full.
Tail—Rather small, and carried low.
Thighs—Strong, long, and well set into body.
Legs—Long, strong, and black or slate in color.
Toes—Four.
Plumage—Close and hard, brilliant metallic black.
Carriage—Very upright, dignified, and watchful.

THE HEN.

Beak—Black, strong, and curved; nostrils arched, broad, and cavernous.
Comb—Double-spiked and branching, standing well up, or the branches inclining a little forward, small.
Head—Long.
Eye—Bright and watchful.
Face—Red, and rather bare.
Deaf-ear—Small and white.
Wattles—Red, small, and neatly rounded.
Neck—Long and straight.
Back—Broad, and tapering towards the tail.
Body—Wide and deep.
Breast—Very broad.
Wings—Large, and well clipped up.
Tail—Small in proportion, but well expanded, and carried upright.
Thighs—Long, and well set into body.
Legs—Long, well boned, black or slaty in color.
Plumage—Brilliant metallic black, close and hard.
Carriage—Upright, dignified, and watchful.

Points

Size	5
Comb	3
Shape	3
Condition	3
Deaf-ear	1
	15

Disqualifications.

Plumage any color but black, presence of crest, feathered legs, deformity of any kind, legs any color but black or dark.

BANTAMS.
GAME BANTAMS.

GENERAL SHAPE AND COLOR.

The same as in the corresponding varieties of Game Fowls.

Points in Game Bantams.

Smallness of Size	2
Color	3
Shape of Head and Neck	2
" Body and Wings	2
" Tail	2
" Thighs, Legs, and Toes	2
Condition	2
	15

Disqualifications.

Cocks above 24 oz. or hens above 20 oz.; adult cocks undubbed, color of legs not uniform in the pen, birds not matching in the pen.

SEBRIGHT BANTAMS.

GENERAL SHAPE—THE COCK.

Comb—Double, square in front, fitting close and straight on the head, the top covered with small points, with a peak behind turning slightly upwards.
Head—Small, round in front, carried well back towards the tail.
Beak—Short, slightly curved.
Eye—Full.
Wattles—Broad, rounded on the lower edge.
Deaf-ear—Flat.
Neck—Neat and taper, quite free from hackle feathers.
Breast—Round, full, and carried prominently forward.
Back—Very short, perfectly free from saddle feathers.
Wings—Ample, the points carried very low, almost touching the ground.
Tail—Square, similar to the hen, free from sickle or curved feathers, the feathers broadest towards the end.
Tail Coverts—Straight, round at the end and lying close to the sides of the tail.
Thighs—Very short.
Legs—Short, slender, and very taper.
Plumage—Close, perfectly hen-feathered.
Carriage—Very upright and strutting.

THE HEN.

Very similar to the cock. The comb and wattles much smaller, and the head neater.

COLOR OF GOLD-LACED SEBRIGHTS.

Head, Face, and Wattles—Rich red.
Deaf-ear—White.
Plumage—Rich golden yellow, every feather laced with rich black, that is, having a narrow, even, well-defined rich black edge all round the feathers; the two colors distinct, and not shading into each other, the lacing of the same width on the sides as on the ends of the feathers.
Legs—Slaty blue.

COLOR OF SILVER-LACED SEBRIGHTS.

Similar to the golden, substituting silvery white for the golden yellow ground color.

Points in Sebrights.

Plumage most evenly and distinctly laced throughout	4
Purity of Ground Color in Silver, and richness and clearness of Ground Color in Golden	2
Comb	2
Tail	1
Smallness	2
Symmetry	2
Condition, and General Appearance	2
	15

Disqualifications.

Cocks weighing more than 20 oz.; hens more than 18 oz.

Cocks having either hackle, saddle or sickle feathers.

Legs of any color except slate blue.

BLACK AND WHITE BANTAMS.

GENERAL SHAPE—THE COCK.

Comb—Double, square in front, close and straight on the head, the top covered with small points, with a peak behind, turning slightly upwards.
Head—Small, round, and carried well back towards the tail.
Beak—Short, slightly curved.
Eye—Prominent.
Deaf-ear—Flat and even on the surface.
Wattles—Broad and thin, rounded on the lower edge.
Neck—Very taper, curving well back, so as to bring the back of the head towards the tail; hackle full and long, flowing well over the shoulders.
Breast—Round, and carried prominently forward.
Back—Very short, saddle feathers long.
Wings—Ample, the points drooping so as nearly to touch the ground, the secondaries slightly expanded.
Tail—Full, expanded, well adorned with long, curving sickle feathers, carried well up toward the back of the head.
Thighs—Short.
Legs—Short, clean, and taper.
Carriage—Very upright, proud, and strutting.

THE HEN.

Comb—Same shape as that of cock, but very much smaller.
Head—Small, round, and neat.
Beak—Small.
Eye—Full and quick.
Deaf-ear—Flat, and even on the surface.
Wattles—Small.
Neck—Short and taper, carried well back.
Breast—Round and prominent.
Back—Short.
Wings—Ample, points drooping.
Tail—Full, expanded, carried rather upright.
Thighs—Short.
Legs—Short, clean and taper.
Carriage—Upright and strutting.

COLOR OF BLACK BANTAMS.

Comb, Face, and Wattles—Rich bright red.
Beak—Dark horn color, or black.
Deaf-ear—Pure white.
Plumage—Rich black throughout.
Legs—Black, or very dark leaden blue.

COLOR OF WHITE BANTAMS.

Comb, Face, and Wattles—Rich scarlet red.
Beak—White.
Deaf-ear—Pure white.
Plumage—Pure white, as free from yellow tinge as possible.
Legs—White, with a slight pink tinge on the back, and betwixt the scales.

Points in Black or White Bantams.

Purity of White or richness of black	3
Smallness	3
Symmetry	3
Comb	2
Deaf-ear	2
Condition, and General Appearance	2
	15

Disqualifications.

Cocks more than 20 oz., or hens more than 18 oz.

Legs of black bantams not black or dark leaden blue.

Legs of white bantams of any other color except white.

TURKEYS.

Head and Face—Very bright and rich in color.
Eyes—Bright and clear.
Body—Long and deep.
Wings—Powerful, and well carried.
Breast—Broad, very long, and perfectly straight.
Thighs—Muscular, straight, and strong.
Legs—Very strong, and perfectly straight.
Plumage—Sound, hard, and glossy.
Color—Rich, the birds matching in the pen.

Points.

Size	6
Symmetry	4
Richness of Color, and Matching in the pen	3
Condition	2
	15

Disqualifications.

Crooked breasts, backs, or legs, or deformity in any part.

DUCKS.
AYLESBURY.

GENERAL SHAPE AND COLOR.

Bill—Long and broad; when viewed sideways, nearly straight from the top of the head to the tip of the bill; of a delicate pale flesh color, perfectly free from black or dark marks.
Head—Long and fine.
Neck—Long, slender, and gracefully curved.
Body—Long and deep.
Back—Long and broad.
Wings—Strong, carried well up, and not drooping.
Tail—Feathers stiff and hard, with hard curled feathers in the drake.
Thighs—Short.
Legs—Short and strong; bright light orange color.
Plumage—Pure white throughout.

Points in Aylesbury Ducks.

Purity of Color and Shape of Bill	8
Size	4
Symmetry	3
Purity of Color in Plumage	3
Condition	2
	15

Disqualifications.

Birds so fat as to be down behind, bills deep yellow, or marked with black, plumage of any color except white.

ROUEN DUCKS.

GENERAL SHAPE AND COLOR—THE DRAKE.

Bill—Long, broad, and rather wider at the tip than at the base; when viewed sideways, nearly straight from the crown of the head to the tip of the bill; the longer the better. Color, greenish yellow, without any other color except the black bean at the tip.
Head—Long and fine; rich lustrous green.
Eye—Dark hazel.
Neck—Long, slender, and neatly curved; color, the same lustrous green as the head, with a distinct white ring on the lower part not quite meeting at the back.
Breast—Broad and deep; the front part very rich purplish brown, or claret color; free from gray feathers, the claret color extending as far as possible towards the legs.
Back—Long; higher part ashy gray mixed with green, becoming a rich, lustrous green on the lower part and rump.
Shoulder Coverts—Gray, finely streaked with waving brown lines.
Wings—Grayish brown, mixed with green, with a broad ribbon mark of purple, with metallic reflections of blue and green, and edged with white; the two colors quite distinct.
" *Flight Feathers*—Dark, dusky brown, quite free from white.
Underpart of Body and Sides—Beautiful gray, becoming lighter gray near the vent, and ending in solid black under the tail.
Tail—Feathers hard and stiff; dark ashy brown, the outer web in old birds edged with white.
Tail Coverts—Curled feathers hard and well curled; black, with very rich purple reflections.
Legs and Feet—Orange, with a tinge of brown.

THE DUCK.

Bill—Broad, long, and somewhat flat; brownish orange, with a dark blotch on the upper part.
Head—Long and fine; deep brown, with two light pale brown stripes on each side from the bill past the eye.
Neck—Long, slender, and neatly curved; light brown, penciled with darker brown, and quite free from the least appearance of a white ring.
Breast, Underpart of Body, and Sides—Grayish brown, each feather marked distinctly with a rich dark brown penciling.
Back—Long; light brown, richly marked with green.
Wings—Grayish brown, mixed with green, with a broad riband mark of rich

purple, edged with white, the two colors distinct.

" *Flight Feathers*—Brown, perfectly free from white.

Tail Coverts—Brown, beautifully penciled with broad distinct penciling of dark greenish brown.

Tail—Light brown, with distinct broad wavy penciling of dark greenish brown.

Legs—Orange, or brown and orange.

Points in Rouen Ducks.

Shape and Color of Bill	3
Size	4
Color of Plumage	3
Symmetry	3
Condition	2
	15

Disqualifications.

Bills clear yellow, dark green, blue or lead color; any white in the flight feathers of either sex; birds so fat as to be down behind.

BLACK EAST INDIAN.

GENERAL SHAPE AND COLOR.

Shape—The entire form remarkably slender, neat, and graceful.

Size—The smaller the better.

Plumage—Rich lustrous black, with a brilliant velvety green tint throughout; perfectly free from white or brown feathers on any part whatever.

Bill of the Drake—Very dark yellowish green, without spot or blemish.

Bill of the Duck—Very dark.

Legs—Dark.

Points in Black East Indian Ducks.

Bill	2
Symmetry, Neatness, and Elegance of Form	3
Richness of Plumage	4
Smallness of Size	4
Condition	2
	15

Disqualifications.

White in any part of the plumage.

CALL DUCKS.

Shape—The entire form very short; round and compact, with very full, round, high forehead, and short broad bill.

Size—The smaller the better.

Color—In the gray variety—bill, legs, and plumage the same as in the Rouen.

" In the white variety—bill bright, clear, unspotted yellow.

Plumage—Pure white.

Legs—Bright orange.

Points in Call Ducks.

Smallness of Size	5
Bill and Stop of the Forehead	2
Symmetry and Compactness of Shape	3
Color of Plumage	3
Condition	2
	15

Disqualifications in Gray Call Ducks.

White ring on the neck of the duck; white flight feathers in either sex.

Disqualifications in White Call Ducks.

Colored feathers in any part of the plumage; bills of any color except yellow.

GEESE.

TOULOUSE.

Carriage—Tall and erect; bodies nearly touching the ground.

Color—Breast and body, light gray; back, dark gray; neck, darker gray than back; wings and belly, shading off to white, though but little white visible.

Bill—Pale flesh color.

Legs and Feet—Deep orange, inclined to red.

EMBDEN.

Plumage—Uniformly pure white.

Bill—Flesh color.

Legs and Feet—Orange.

Points in Geese.

Size and Weight	6
Symmetry	4
Color	3
Condition	2
	15

www.ingramcontent.com/pod-product-compliance
Lightning Source LLC
Chambersburg PA
CBHW062319220526
45469CB00008B/2567